WITHDRAWN

ELEMENTS OF COMPUTATIONAL HYDRAULICS

Christopher G. Koutitas
*Professor of Coastal Engineering,
Democritos University of Thrace, Greece*

PENTECH PRESS
London: Plymouth

Distributed in the USA by
Chapman and Hall
New York

First published 1983 by
Pentech Press Limited
Estover Road, Plymouth
Devon

© Pentech Press Limited, 1983

British Library Cataloguing in Publication Data
Koutitas, C. G.
  Elements of computational hydraulics.
  1. Hydraulic engineering—Data processing
  I. Title
  627'.028'54   TC157.8
  ISBN 0-7273-0503-4

Exclusive distributor USA
Chapman & Hall
733 Third Avenue,
New York, NY 10017

Library of Congress Cataloging in Publication Data
Koutitas, Christopher G.
  Elements of computational hydraulics.
  Bibliography: P.
  Includes index.
  1. Hydraulics—Data processing.   I. Title.
  TC157.8.K68   1982   627'.028'54   82-22154
  ISBN 0-412-00361-9

Filmset by Mid-County Press, London SW15
Printed in Great Britain by
Billing & Sons Limited, London and Worcester

Preface

This volume is primarily intended for advanced undergraduate students of civil engineering and also for practicing engineers interested in this field. A basic course in applied (engineering) mathematics including some elements of numerical analysis, an introductory course in hydraulics, and some experience in FORTRAN programming are considered as the necessary prerequisites to follow easily the material presented in this volume.

The book is at an introductory level throughout as its aim is to facilitate the engineer, having a separate knowledge and experience of hydraulics and computer programming, to successfully match the two and to realise that he can effectively use the computer to solve problems in hydraulics.

It is by no means to be assumed that the book contains all the information needed for all aspects of research and practical applications. Further study in numerical analysis, hydraulics and computer programming is required in order to master more complex situations, but it is hoped that the reader will manage to achieve some insight into the vast possibilities offered by the digital computer.

The material is presented in the following order. Some basic notions of numerical analysis are presented first. The techniques of numerical interpolation and integration, numerical expansion in Fourier series and the numerical solution of systems of algebraic equations are comprehensively developed. The method of finite differences is presented and applied to some basic differential equations occurring in hydraulics.

The remaining chapters are concerned with the numerical treatment of mathematical models in the most fundamental sectors of hydraulics, i.e. the flow in closed and open conduits, the flow in porous media and diffusion–dispersion problems. The last chapter is an introduction to the method of finite elements, a numerical method attracting considerable interest in contemporary engineering. The introduction to the method is performed in an elementary manner with repetitive applications to simple problems.

The applications included in each chapter are accompanied by complete computer programs in FORTRAN. The applications and the corresponding programs are simple as it is intended to keep to an introductory level. It is a matter of further research and practice for the reader to optimize and generalise the computer programs herein contained.

C. G. Koutitas

Contents

1. **ELEMENTS OF NUMERICAL ANALYSIS**   1
   1.1. Introduction   1
   1.2. Definitions — General concepts   1
   1.3. Numerical approximation and interpolation   3
   1.4. Numerical integration   8
   1.5. Solution of systems of algebraic equations   11
   1.6. Numerical analysis in finite Fourier series   18
   1.7. Method of finite differences in the solution of differential equations   21

2. **NUMERICAL SOLUTION OF PARTIAL DIFFERENTIAL EQUATIONS COMMON IN HYDRAULICS**   27
   2.1. Form and occurrence of some partial differential equations   27
   2.2. Numerical solution of parabolic equations   29
   2.3. The hyperbolic equation and the method of characteristics   35
   2.4. Elliptic equations   43

3. **FLOW IN CLOSED CONDUITS**   48
   3.1. Mathematical models for steady flow in pipes and pipe networks   48
   3.2. Steady flow in pipe networks (the Hardy-Cross method)   51
   3.3. Non steady flow. Water hammer   55

4. **OPEN CHANNEL FLOW**   65
   4.1. Mathematical models for non-steady flow in open channels   65
   4.2. Numerical solutions for long wave propagation   67
   4.3. Steady non-uniform flow. Backwater curve analysis   80

5. **GROUNDWATER FLOW**   88
   5.1. Mathematical models for flow in porous media   88
   5.2. Application of mathematical models to flow in porous media   90

6. **ADVECTIVE DIFFUSION AND DISPERSION**   99
   6.1. Mathematical models for diffusion–dispersion of matter   99
   6.2. Numerical solution of mathematical models of diffusion and dispersion   101

7. THE METHOD OF FINITE ELEMENTS 105
7.1. Historical background — an introduction to the method 105
7.2. The Ritz and Galerkin methods for the approximate solution of differential equations 106
7.3. Discretisation by finite element shape functions 113
7.4. Derivation of finite element equations by the Ritz or Galerkin method 116
7.5. Application to long linear waves in open channels 126
7.6. Application to groundwater flow 131

REFERENCES 136

INDEX 137

To Mary,
Maria Christina,
and George

1

Elements of numerical analysis

1.1. INTRODUCTION

Instruction in theoretical and applied hydraulics aims at supplying the engineer with the necessary knowledge for the correct design and construction of hydraulic works. An understanding and quantitative description of the relevant physical phenomena which permits the optimal design of the several components of a hydraulic works requires sophisticated computational procedures, especially when the investigation and design is either aided or totally accomplished by the use of mathematical models.

These computational procedures become ever more extensive as the problems investigated and the assignments undertaken become more complex and their execution by hand or with the aid of simple calculators becomes time consuming or even impossible.

The computational problems arising in hydraulics are characterised by their wide variety and most of them can be classified in the wider class of problems of continuum mechanics.

The rapidly increasing access to and use of electronic digital computers and the recent introduction of microcomputers has permitted, as in other branches of engineering, the application of computers to hydraulics. The computer is, of course, by no means a panacea for all the computational difficulties which may arise. Its effective use assumes a sound knowledge of numerical analysis and programming and their combined application to the mathematical problem in hand.

1.2. DEFINITIONS — GENERAL CONCEPTS

The easiest way, nowadays, to investigate a physical phenomenon, when an electronic computer is available, is through the synthesis and solution of a mathematical model. A mathematical model is composed of the mathematical expressions quantifying fundamental physical principles, such as the equilibrium of forces, the conservation of energy and mass, etc. These mathematical expressions are adapted

2 ELEMENTS OF NUMERICAL ANALYSIS

and simplified in each case according to the special features of the problem to be tackled, i.e. boundary conditions, the range of values of the intermediate parameters, etc.

For the numerical solution of mathematical models describing physical phenomena, algorithms are formulated and used. An algorithm can be defined as a computational method or more simply an ensemble of rules determining the order and the form of the numerical operations to be applied to a set of data $\mathbf{a}(a_1 \ldots a_n)$ in order to find a new set of values $\mathbf{x}(x_1 \ldots x_m)$ forming the solution of the problem.

An algorithmic procedure can be concisely represented by

$$\mathbf{x} = f(\mathbf{a}) \tag{1.1}$$

From a mathematical standpoint the main preoccupation is to determine whether the computational method giving $\mathbf{x}$ from the application of f on $\mathbf{a}$ is well posed or not. For an algorithm to be 'well posed' it must satisfy the following conditions:

(1) A solution exists for a given $\mathbf{a}$ (it may not be well posed, for example, if a computational method for the solution of a system of simultaneous equations through successive iterations does not converge).

(2) The computation must lead to a single solution $\mathbf{x}$ for a given $\mathbf{a}$.

(3) The results, $\mathbf{x}$, must be connected to the data, $\mathbf{a}$, through the well known Lipschitz relation. If

$$|\delta \mathbf{a}| \leqslant \eta, \quad \text{then} \quad |\delta \mathbf{x}| \leqslant M |\delta \mathbf{a}| \tag{1.2}$$

where M is a bounded natural number, $M = M(\mathbf{a}, n)$.

It is mentioned without proof that the application of the third criterion to an elliptic problem shows that it is well posed only as a boundary value problem and not as an initial value one.

Of the computational schemes included in most treatises on numerical analysis the most common in hydraulics are numerical interpolation and integration, analysis by finite Fourier series, numerical solutions of algebraic systems and numerical solutions of differential equations (by either finite differences or finite elements).

The following sections will be focused on algorithms and some numerical schemes will be subsequently presented for the integration of the differential equations most common in hydraulics.

1.3. NUMERICAL APPROXIMATION AND INTERPOLATION

Assume that for a function $f(x)$ a series of x_i values and the corresponding $f_i (i = 1 \ldots n)$ are given. The interpolation is defined as the approximate computation of the f value for a $x \neq x_i$. Such series of pairs f_i, x_i may arise from laboratory experiments or may be given in tables describing the use of an instrument or a nomograph. When uncontrollable errors are included in the f_i values (as with laboratory experiments), then the curve passing from all the points on the f–x plane may not be smooth and the 'cloud' of points is usually approximated by a curve passing among the points in a weighted way. This is usually done by the least-squares method, achieving a minimisation of the variance between the smooth curve and the experimental points.

The least-squares method of approximation is illustrated through the following example:

EXAMPLE 1.1

A series of velocity values at different depths on a cross section of an open channel is given in semi-logarithmic scale. If z is the distance from the bottom (in m) and u the water velocity (in cm/s) the data are (Table 1.1):

Table 1.1

$x = \ln z$	0	0.2	0.4	0.6	0.8	1	1.2	1.4	1.6	1.8	2.0
$f = u$	20	30	40	50	50	60	70	80	90	95	100

It is recalled that in uniform turbulent open channel flow the following velocity distribution holds with sufficient accuracy:

$$u = \frac{u^*}{K} \ln\left(\frac{z}{z_0}\right) \tag{1.3}$$

where K is the von Kármán constant ($K = 0.4$), z_0 the roughness height (a measure of the size of bottom irregularities) and u^* the friction velocity, a function of the bottom shear stress τ_0,

$$u^* = \sqrt{\tau_0/p} \tag{1.4}$$

What is the velocity at a depth 3.5 m?

4 ELEMENTS OF NUMERICAL ANALYSIS

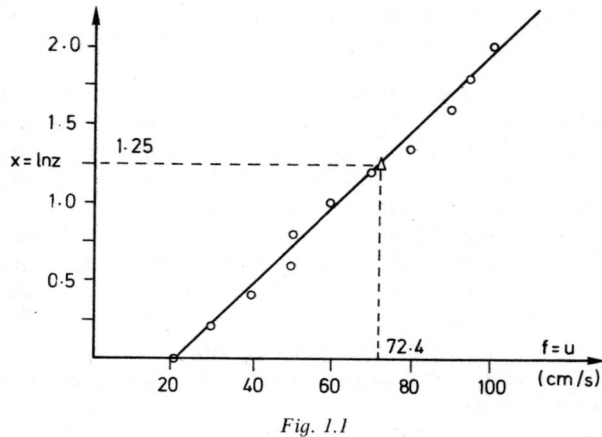

Fig. 1.1

From the graphic representation of the spread of given points (Fig. 1.1) and according to the relation 1.3 the most appropriate curve approximating the data is a straight line. Its mathematical expression is

$$f = \alpha x + \beta \tag{1.5}$$

The method of least squares requires for the computation of α and β the minimisation of the sum of the squares of the deviations, $f - f_i$.

$$R = \sum_{i=1}^{n} (f - f_i)^2 = \sum_{i=1}^{n} (\alpha x_i + \beta - f_i)^2 = \min \tag{1.6}$$

The minimisation in Equation 1.6 is achieved by making the derivatives zero,

$$\frac{\partial R}{\partial \alpha} = 0 \tag{1.7}$$

$$\frac{\partial R}{\partial \beta} = 0 \tag{1.8}$$

leading to the following expressions for α and β:

$$\alpha = \frac{n \sum x_i f_i - \sum x_i \sum f_i}{n \sum x_i^2 - (\sum x_i)^2} \tag{1.9}$$

$$\beta = \frac{\sum x_i^2 \cdot \sum f_i - \sum x_i \sum x_i f_i}{n \sum x_i^2 - (\sum x_i)^2} \tag{1.10}$$

For a given n (the number of pairs of values, $n=11$) and given pairs x_i, f_i the computation of α, β can be programmed in FORTRAN as follows:

```
C      LEAST SQUARE CURVE FITTING
       DIMENSION X(100), F(100)
       READ (5,5) N, (X(I), F(I), I=1,N)
5      FORMAT (I4/(10F 7.0))
       SUM1=0.
       SUM2=0.
       SUM3=0.
       SUM4=0.
       DO 7 I=1,N
       SUM1=SUM1+X(I)
       SUM2=SUM2+F(I)
       SUM3=SUM3+X(I)**2
7      SUM4=SUM4+X(I)*F(I)
       A=(SUM4*N-SUM1*SUM2)/(N*SUM3-SUM1**2)
       B=(SUM3*SUM2-SUM1*SUM4)/(N*SUM3-SUM1**2)
       WRITE (6,8) A, B
8      FORMAT (2F 10.4)
       STOP
       END
```

For the given values of x_i, f_i the computed α, β values are $\alpha = 40.45$, $\beta = 21.28$ and for $\ln z = \ln(3.5) = 1.25 = x$, it is found that $f = u = 72.4$ cm/s.

When no errors are included in the f_i values the computation of f for a given x, i.e. the numerical interpolation, can be done in two ways depending on the distance h between the abscissae x_i. For constant h ($h = \Delta x = x_{i+1} - x_i$, for all i), the Newton formula can be applied (in its forward or backward form). If the forward difference operator is symbolised by Δ,

$$\Delta f(x) = f(x+h) - f(x) \tag{1.11}$$

the backward difference operator by ∇,

$$\nabla f(x) = f(x) - f(x-h) \tag{1.12}$$

and the translation operator by E,

$$E f(x) = f(x+h) \tag{1.13}$$

where

$$E^k f(x) = f(x+kh)$$

6 ELEMENTS OF NUMERICAL ANALYSIS

Table 1.2

x	f	Δ	Δ²	Δ³
0	0			
1	3	3	2	
2	8	5	14	12
3	27	19	18	4
4	64	37	24	6
5	125	61		

then the following symbolic equation holds:

$$\Delta^k = (E-1)^k \tag{1.14}$$

The Newton formulae is derived from the expansion in Taylor series. If the $f(x+ah)$ and $f(x-ah)$ values are required, with $0 \leqslant a \leqslant 1$, then the Newton formulae are:

Forward formula,

$$f(x+ah) = \left(1 + a\Delta + \frac{a(a-1)}{2!}\Delta^2 + \cdots\right)f(x) + R \tag{1.15}$$

Backward formula,

$$f(x-ah) = \left(1 + \nabla + \frac{a(a-1)}{2!}\nabla^2 + \cdots\right)f(x) + R \tag{1.16}$$

The truncation error R due to the interruption of the Taylor series is related to the order of the last included term; if n terms are retained then

$$R \propto h^{n+1} f^{(n+1)}(\xi), \quad x_0 < \xi < x_n \tag{1.17}$$

where the upper index in parenthesis $(n+1)$ denotes the order of the derivative.

The Newton formulae are applied in the next example.

EXAMPLE 1.2

The pairs x_i, f_i with $h = \Delta x = 1$ are given (Table 1.2) and $f(x = 2.5)$ is required. From the pairs f_i, x_i the differences $\Delta, \Delta^2, \Delta^3 \ldots$ are computed and tabulated below. Application of Equation 1.15 for $x = 2.0$, $a = 0.5$ and $n = 3$ (the largest possible) gives the $f(2.5)$ value

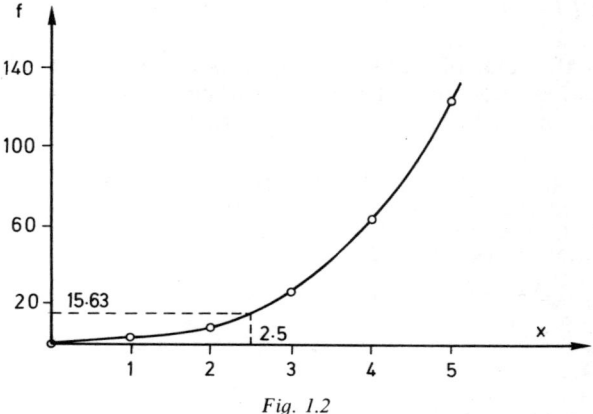

Fig. 1.2

$$f(2.5) = 8 + 0.5 \times 19 + \frac{0.5 \times (0.5-1)}{2} \times 18 +$$

$$+ \frac{0.5 \times (0.5-1)(0.5-2)}{2 \times 3} \times 6 = 15.63$$

A graphical representation of the function is given in Fig. 1.2.

A more general procedure applicable independently of the distance between the x_i, x_{i+1} pairs is the method of Lagrange polynomials. For n given pairs (x_i, f_i) a polynomial of $n-1$ degree passing through all points in the f–x plane is constructed. The general form of the polynomial is

$$\Phi(x) = \sum_{i=1}^{n} a_i \prod_{\substack{k=1 \\ k \neq i}}^{n} (x - x_k) \qquad (1.18)$$

where Π indicates the product of $x - x_k$ terms. The computation of a_i coefficients is done on the basis that $\Phi(x_i) = f_i$ for all i. It is found that,

$$a_i = \frac{f_i}{\prod_{\substack{k=1 \\ k \neq i}}^{n} (x_i - x_k)} \qquad (1.19)$$

The method is illustrated in the following example.

EXAMPLE 1.3

Compute the value of $f(2.5)$ for the given six pairs (x_i, f_i) in Example 1.2. Equations 1.18 and 1.19 are programmed with $n=6$. The fifth-degree polynomial giving $f(2.5)$ is constructed.

```
C     LAGRANGE POLYNOMIALS INTERPOLATION
      DIMENSION X(100), F(100), A(100)
      READ (5,5)N, (X(I), F(I), I=1,N)
   5  FORMAT (I4/(10F 7.0))
      READ (5,8) XC
   8  FORMAT (F 7.0)
      DO 6 I=1,N
      P=1.
      DO 3 K=1,N
      IF (K-I) 13,3,13
  13  P=P*(X(I)-X(K))
   3  CONTINUE
      A(I)=F(I)/P
   6  CONTINUE
      FF=0.
      DO 16 I=1,N
      PP=1.
      DO 7 K=1,N
      IF(K-I) 15,7,15
  15  PP=PP* (XC-X(K))
   7  CONTINUE
      FF=FF+A(I)*PP
  16  CONTINUE
      WRITE(6,19) XC, FF
  19  FORMAT (2F 10.4)
      STOP
      END
```

For XC=2.5 it is found that FF=15.43.

1.4. NUMERICAL INTEGRATION

Numerical integration is defined as the approximate numerical evaluation of the integral

$$I = \int_{x_0}^{x_1 = x_0 + h} f(x)\,dx \qquad (1.20)$$

Several computational schemes for numerical integration exist and can be derived from Newton's formula. Their form depends on the number of terms retained in the series.

Assuming the approximation of $f(x)$ by $\varphi(a)$, for example,

$$\varphi(a) = (1 + a\Delta)f_0 \tag{1.21}$$

the integral becomes

$$I = \int_{x_0}^{x_1} f \, dx = h \int_0^1 \varphi(a) \, da = h \int_0^1 (1 + a\Delta)f_0 \, da$$

$$= h\left(a + a^2 \frac{\Delta}{2}\right)f_0 \bigg|_0^1 = \left(f_0 + \left(\frac{f_1 - f_0}{2}\right) \times 1\right)h$$

$$= \left(\frac{f_1 + f_0}{2}\right)h \tag{1.22}$$

which is, of course, the trapezoidal rule.

The best known method for the synthesis of numerical integration schemes is the method of undetermined coefficients. All the classical formulae for numerical integration derive as special cases of this method. Using this method the integral is approximated by the sum:

$$I = \sum_{k=1}^{n} H_k f(x_k) \tag{1.23}$$

If the end points are included in the x_k array the formula is called closed, otherwise it is called open. The number of the unknown coefficients H_k determines the maximum degree of the polynomials that can be integrated through Equation 1.23 without error. This actually forms the procedure for the determination of the H_k's, i.e. through the equality of the right hand side of Equation 1.23 and the integrals of polynomials of degree $0, 1, 2, \ldots n-1$.

Take for example the formula

$$I = \int_0^1 f \, dx = H_0 f_0 + H_1 f_1 \tag{1.24}$$

For the computation of H_0, H_1 the polynomials $f = 1$, $f = x$ are used.

$$\int_0^1 1 \, dx = 1 = H_0 \times 1 + H_1 \times 1 \tag{1.25}$$

10 ELEMENTS OF NUMERICAL ANALYSIS

Table 1.3

Month	1	2	3	4	5	6	7	8	9	10	11	12
Discharge × 10^6 m³/day	8	6	5	5	4	1	1	1	5	6	7	7

$$\int_0^1 x \, dx = \tfrac{1}{2} = H_0 \times 0 + H_1 \times 1 \qquad (1.26)$$

giving $H_0 = 1/2$, $H_1 = 1/2$, which is the trapezoidal rule.

The well known Simpson formula can be obtained through the computation of H_0, H_1, H_2, in

$$I = H_0 f_0 + H_1 f_{1/2} + H_1 f_1 = \int_0^1 f \, dx \qquad (1.27)$$

For $f = 1, x, x^2$ it is found that:

$$I = \tfrac{1}{6}(f_0 + 4f_{1/2} + f_1) \qquad (1.28)$$

The truncation error introduced can be computed from the integration of a polynomial of degree n.

EXAMPLE 1.4

Compute, by means of the trapezoidal rule, the volume of water collected during one year in a reservoir upstream of a dam built on a river where the discharges (measured every month) are given in Table 1.3.
What must be the storage capacity of the reservoir to meet the need for a continuous water supply of 42 m³/s?
For the computation of the reservoir storage capacity the curve of accumulated volume with time is evaluated numerically. The integration is performed with $h = 30$ days, for all months, and the results of the integration are printed continuously.
The FORTRAN listing can have the form:

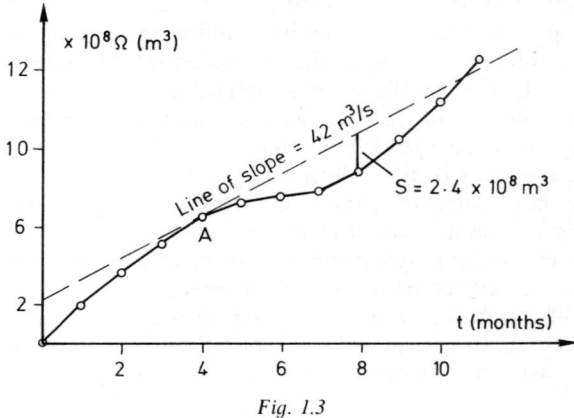

Fig. 1.3

```
C     NUMERICAL INTEGRATION
      DIMENSION F (100)
      READ (5,5)N, H, (F(I), I=1,N)
5     FORMAT (I4,F 7.0/(10F 7.0))
      SUM=0.
      NN=N-1
      DO 10 I=1,NN
      SUM=SUM+(F(I)+F(I+1))/2.*H
      WRITE (6,15) SUM
15    FORMAT (E14.6)
10    CONTINUE
      STOP
      END
```

From the values of the integral SUM at the end of each month the curve of accumulated volume versus time is traced. The distance between the curve and a straight line of slope 42 m^3/s (tangent at the high point A) is measured. The distance, equal to 2.4×10^8 m^3 is the necessary storage capacity of the reservoir (see Fig. 1.3).

1.5. SOLUTION OF SYSTEMS OF ALGEBRAIC EQUATIONS

Many hydraulic computations end up as solutions of simultaneous algebraic equations. The most common cases, as it will be shown later, refer to the numerical solutions of differential equations involved in the mathematical models of steady or transient flows. Special computational schemes, according to the form of the matrices, have been developed. Some general characteristics of the most common algorithms will be presented here.

12 ELEMENTS OF NUMERICAL ANALYSIS

The solution of systems of algebraic equations can be done either directly or through successive iterations (i.e. successive approximations converging to the real solution). The criteria for the selection of the first or the second method are:

(1) The amount of arithmetic computation and the storage capacity of the computer available.
(2) The accuracy of the solution required.
(3) The possibility of controlling the accuracy of the results obtained by the numerical solution.

The direct solution algorithms lead to results by means of a finite predictable number of arithmetic operations but at a risk of intrusion and amplification of errors. The iterational algorithms lead to convergence to the exact solution but through an unpredictable number of arithmetic operations.

1.5.1. Direct solution algorithms

1.5.1.1. *The Gauss–Jordan elimination*

The matrix of the coefficients of the unknowns of the system

$$[A]\{x\} = \{b\} \tag{1.29}$$

is transformed through successive eliminations to an upper triangular matrix and the solution is found by back substitution

$$[A] \rightarrow [U], \quad \{b\} \rightarrow \{c\}, \quad [U]\{x\} = \{c\} \tag{1.30}$$

where $[U]$ is upper triangular matrix. The equations are decoupled, the last one is solved first and the solution continues from x_n to $x_{n-1} \ldots x_1$.

For a given matrix $[A]$ and vector $\{b\}$ the method of Gauss–Jordan can be programmed as follows:

```
C     SYSTEM SOLUTION BY GAUSS ELIMINATION
      DIMENSION A(50,50), X(50), B(50), C(50,50)
      READ(5,5)N, ((A(I,J), I=1,N), J=1,N)
    5 FORMAT (I4/(10F 7.0))
      READ (5,6) (B(I), I=1,N)
    6 FORMAT (10F 7.0)
      DO 3 I=1,N
    3 X(I)=0.
      NN=N−1
      DO 20 I=1,N
      DO 20 J=1,N
```

```
20    C(I,J)=A(I,J)
      DO 4 K=1,NN
      L=K+1
      DO 15 I=L,N
      DO 16 J=1,N
16    C(I,J)=A(I,J)-A(K,J)*A(I,K)/A(K,K)
      B(I)=B(I)-B(K)*A(I,K)/A(K,K)
15    CONTINUE
      DO 18 I=1,N
      DO 18 J=1,N
18    A(I,J)=C(I,J)
4     CONTINUE
      DO 11 I=1,N
11    WRITE(6,10) (A(I,J), J=1,N)
      WRITE(6,10) (B(I), I=1,N)
      DO 7 K=N, 1, -1
      SUM=0.
      DO 8 I=1,N
      IF(I.EQ.K) GO TO 8
      SUM=SUM+A(K,I)*X(I)
8     CONTINUE
      X(K)=(B(K)-SUM)/A(K,K)
7     CONTINUE
      WRITE(6,10) (X(I), I=1,N)
10    FORMAT(10F10.3)
      STOP
      END
```

EXAMPLE 1.5

Compute the unknown quantities $x(x_1 \ldots x_5)$ satisfying the system of equations

$$[A]\{x\} = \{B\}$$

$$[A] = \begin{bmatrix} 4 & 1.5 & 0.7 & 1.2 & 0.5 \\ 1 & 6 & 0.9 & 1.4 & 0.7 \\ 0.5 & 1 & 3.9 & 3.2 & 0.9 \\ 0.2 & 2 & 0.2 & 7.5 & 1.9 \\ 1.7 & 0.9 & 1.2 & 2.3 & 4.9 \end{bmatrix} \quad \{B\} = \begin{Bmatrix} 5 \\ 6 \\ 7 \\ 8 \\ 9 \end{Bmatrix}$$

by Gauss–Jordan elimination.

The application of the previous program with $n=5$ and the fixed values of $[A]$ and $\{B\}$ give for the unknown $\{x\}$

$$\{x\} = [0.589 \quad 0.521 \quad 0.831 \quad 0.617 \quad 1.076]^T$$

1.5.1.2. Choleski–Banachiewicz method

This method is based on the analysis of the coefficients matrix $[A]$ to a product of two matrices a lower triangular $[L]$ and an upper triangular $[U]$

$$[A] = [L][U] \tag{1.31}$$

The solution with respect to an intermediate variable $\{y\}$ of the system

$$[L]\{y\} = \{f\} \tag{1.32}$$

is achieved by direct substitution, and the solution ofr the $\{x\}$ variable is subsequently achieved by back substitution

$$[U]\{x\} = \{y\} \tag{2.33}$$

In the case of a tridiagonal matrix $[A]$, which usually occurs during the integration of parabolic and hyperbolic partial differential equations, $[A]$ can be expressed as the product of triangular matrices according to Thomas's method, as follows:

$$[A] = \begin{bmatrix} b_1 & c_1 & & \\ \alpha_2 & b_2 & c_2 & 0 \\ & \alpha_3 & b_3 & c_3 \\ 0 & & & \end{bmatrix} = \begin{bmatrix} a'_1 & & & \\ a_2 & a'_2 & & 0 \\ & a_3 & a'_3 & \\ 0 & & & \end{bmatrix} \begin{bmatrix} 1 & b'_1 & & 0 \\ & 1 & b'_2 & \\ & & 1 & b'_3 \\ & & & 0 \end{bmatrix}$$

$$\tag{1.34}$$

where

$$a'_i = b_i - b'_{i-1} a_i \tag{1.35}$$

$$b'_i = c_i / a'_i \tag{1.36}$$

The solution with respect to the auxiliary variable $\{y\}$ has the form

$$y_i = \frac{f_i - a_i y_{i-1}}{a'_i} \tag{1.37}$$

The unknowns $\{x\}$ are finally calculated through the relations

$$\begin{aligned} x_n &= y_n \\ x_i &= y_i - b'_i x_{i+1} \end{aligned} \tag{1.38}$$

1.5.1.3. *Solution by successive iterations*

The solution by successive iterations is performed by expressing the matrix $[A]$ as the sum

$$[A] = [B] + [C] \rightarrow [B]\{x\} + [C]\{x\} = \{\vec{f}\} \rightarrow [B]\{x\} = -[C]\{x\} + \{\vec{f}\}$$

(1.39)

If the upper index in parenthesis (v) denotes the order of iteration the method actually consists of a succession of computations of new values $\{x\}^{(v+1)}$ from previous values $\{x\}^{(v)}$ by means of the relation

$$[B]\{x\}^{(v+1)} = -[C]\{x\}^{(v)} + \{\vec{f}\} \qquad (1.40)$$

or

$$\{\vec{x}\}^{(v+1)} = [Q]\{\vec{x}\}^{(v)} + [B^{-1}]\{\vec{f}\} \qquad (1.41)$$

where

$$[Q] = -[B^{-1}][C] \qquad (1.42)$$

By induction it is found that

$$\{x\}^{(v+1)} = [Q^{v+1}]\{x\}^{(0)} + [I + Q + \cdots Q^v][B^{-1}]\{f\} \qquad (1.43)$$

Convergence to the exact solution is ensured if $[Q]^{v+1} \rightarrow 0$ with increasing v.

From the theory of matrices it can be proved that a necessary condition for such convergence is that the spectral radius* of the $[Q]$ matrix (maximum eigenvalue) be less than unity. The rate of convergence depends on the decomposition of $[A]$ into the sum $[B] + [C]$. The smaller the spectral radius of $[Q]$ the higher is that rate[14,21]. Three specific forms of the iterative method with wide use will be described.

1.5.1.4. *Method of Gauss–Jacobi*

The $[A]$ matrix according to this method is represented as the sum

$$[A] = [L] + [D] + [U] \qquad (1.44)$$

---

* The spectral radius is defined as the maximum characteristic value of the matrix.

where

$$[L] = \begin{bmatrix} 0 & & & 0 \\ & 0 & & \\ & & 0 & \\ a_{ik} & & & \end{bmatrix} \quad (1.45)$$

$$[D] = \begin{bmatrix} a_{11} & & & 0 \\ & a_{12} & & \\ 0 & & & \end{bmatrix} \quad (1.46)$$

$$[U] = \begin{bmatrix} 0 & & & \\ & 0 & a_{ik} & \\ & 0 & & \end{bmatrix} \quad (1.47)$$

Equation 1.29 becomes

$$[L]\{x\}^{(v)} + [D]\{x\}^{(v+1)} + [U]\{x\}^{(v)} = \{f\} \quad (1.48)$$

or

$$\{x\}^{(v+1)} = -[D]^{-1}[L+U]\{x\}^{(v)} + [D]^{-1}\{f\} \quad (1.49)$$

The convergence depends on the value of the spectral radius of $[D^{-1}][L+U]$ satisfying the inequality,

$$\max_{\substack{j \\ j \neq i}} \sum \left|\frac{a_{ij}}{a_{ii}}\right| \leq 1 \quad (1.50)$$

for all i ($i = 1, 2, \ldots n$).

1.5.1.5. *The method of Gauss–Seidel*

The matrix $[A]$ is expressed as previously but multiplying the $[L]$ matrix by the most recent $\{x\}^{(v+1)}$ values. Thus Equation 1.49 becomes

$$\{x\}^{(v+1)} = -[U][L+D]^{-1}\{x\}^{(v)} + [L+D]^{-1}\{f\} \quad (1.51)$$

The convergence criterion is the same, i.e. the dominancy of the diagonal terms a_{ii} over the sum of the off-diagonal coefficients in the same equation (satisfying inequality 1.50).

1.5.1.6. Relaxation methods

In this method the convergence of the solution is accelerated by the introduction of a parameter ω which is estimated as follows. Let us consider the decomposition of the matrix $[A]$, i.e.

$$[A] = [L] + [D] + [U] = [L+I] + [U+D-I] \quad (1.52)$$

and

$$[L+I]\{x\}^{(v+1)} + [U+D-I]\{x\}^{(v)} = \{f\} \quad (1.53)$$

Solving for $\{x\}^{(v+1)}$, Equation 1.53 becomes,

$$\{x\}^{(v+1)} = \{x\}^{(v)} - \omega([L]\{x\}^{(v+1)} + [U+D]\{x\}^{(v)} - \{f\}) \quad (1.54)$$

or

$$\{x\}^{(v+1)} = [B]\{x\}^{(v)} + \{\vec{f}\} \quad (1.55)$$

where

$$[B] = [I + \omega L]^{-1}[I - \omega D - \omega U] \quad (1.56)$$

The spectral radius ρ of B is estimated and its minimisation gives for ω,

$$\omega_{opt} = \frac{2}{1 + \sqrt{1-\rho^2}} \quad (1.57)$$

In the following example, the Gauss–Seidel method is applied to the system of equations of Example 1.5.

EXAMPLE 1.6

Solve with successive iterations the system of Example 1.5, and compare the solution vectors. The convergence criterion to be taken equal to 10^{-4}.

$$\max|(x_i^{(v+1)} - x_i^{(v)})| \leqslant 10^{-4}$$

Equation 1.51 is used and the procedure is programmed as follows:

```
C     SYSTEM SOLUTION BY GAUSS SEIDEL ITERATION
      DIMENSION A(50,50), B(50), X(50)
      READ(5,1) N, TEST
    1 FORMAT (I4, F10.0)
      READ(5,2)((A(I,J), I=1,N), J=1,N)
    2 FORMAT (10F 7.0)
      READ(5,2)(B(I), I=1,N)
      DO 3 I=1,N
```

18   ELEMENTS OF NUMERICAL ANALYSIS

```
    3   X(I)=0.
        ITER=0
100     DIFMX=0.
        ITER=ITER+1
        IF(ITER.GT.100) STOP 1
        DO 4 I=1,N
        TEMP=X(I)
        SUM=0.
        DO 5 J=1,N
        IF(J.EQ.I) GO TO 5
        SUM=SUM+X(J)*A(I,J)
    5   CONTINUE
        X(I)=(B(I)-SUM)/A(I,I)
        DIF=ABS (TEMP-X(I))
        IF(DIF.GT.DIFMX) DIFMX=DIF
    4   CONTINUE
        IF(DIFMX.GT.TEST) GO TO 100
        WRITE(6,11)(X(I), I=1,N)
   11   FORMAT(10F10.4)
        STOP
        END
```

With the use of matrices $[A]$, $\{B\}$ of Example 2.5 and with TEST $= 10^{-4}$, the solution is

$$\{x\} = [0.589 \quad 0.521 \quad 0.831 \quad 0.617 \quad 1.076]^T$$

1.6. NUMERICAL ANALYSIS IN FINITE FOURIER SERIES

Processing hydrologic data and hydraulic measurements as the discharge of rivers, water level variations and velocity components in lakes and coastal region usually requires analysis by Fourier series with the consequent estimation of the amplitude of several harmonic components included in the time series. This analysis is based on the orthogonality of the coordinate functions 1, cos x, cos $2x$, cos $3x$ ..., and sin x, sin $2x$ ....

As the field variables are usually stored and used in discrete form and not as continuous functions, the detection of the harmonic components is done in an arithmetic way on the basis that the above mentioned sets of base functions, cos and sin, are orthogonal even in a numerical sense. This property permits the numerical analysis in finite Fourier series which will be described here.

Assume $2N$ measurements of a variable $f_i(x)$ for equally spaced points x_i. If the length on the x axis is L, the abscissa x_i is expressed as:

$$x_i = \frac{Li}{2N}, \quad i=0, \ldots 2N-1 \qquad (1.58)$$

ELEMENTS OF NUMERICAL ANALYSIS 19

The harmonic component of the highest frequency that can be included in the analysis in Fourier series of f in discrete form has a wave length equal to $2\Delta x = 2L/(2N-1)$, which is the finest possible resolution. Therefore, the $f(x)$ function can be approximated by

$$f(x) = \frac{A_0}{2} + \sum_{k=1}^{N-1} \left(A_k \cos\left(\frac{2\pi k x}{L}\right) + B_k \sin\left(\frac{2\pi k x}{L}\right) + \right.$$
$$\left. + \frac{A_N}{2} \cos\left(\frac{2\pi N x}{L}\right) \right) \quad (1.59)$$

The computation of the coefficients $A_0 \ldots A_N, B_1 \ldots B_{N-1}$ is based on the orthogonality of the functions cos, sin in a numerical sense, i.e.

$$\sum_{p=0}^{2N-1} \cos\left(\frac{2\pi k\, Lp}{L\, 2N}\right) \cos\left(\frac{2\pi m\, Lp}{L\, 2N}\right) = \begin{matrix} 0 \\ N \\ 2N \end{matrix} \begin{cases} k \neq m \\ k = m \neq 0, N \\ k = m = 0, N \end{cases}$$
$$(1.60)$$

$$\sum_{p=0}^{2N-1} \sin\left(\frac{2\pi k\, Lp}{L\, 2N}\right) \sin\left(\frac{2\pi m\, Lp}{L\, 2N}\right) = \begin{matrix} 0 \\ N \\ 0 \end{matrix} \begin{cases} k \neq m \\ k = m \neq 0, N \\ k = m = 0, N \end{cases} \quad (1.61)$$

$$\sum_{p=0}^{2N-1} \cos\left(\frac{2\pi k\, Lp}{L\, 2N}\right) \sin\left(\frac{2\pi m\, Lp}{L\, 2N}\right) \equiv 0 \quad (1.62)$$

The solution for A_k, B_k is,

$$A_k = \frac{1}{N} \sum_{p=0}^{2N-1} f(x_p) \cos\left(\frac{2\pi k x_p}{L}\right), \quad k = 0, \ldots N \quad (1.63)$$

$$B_k = \frac{1}{N} \sum_{p=0}^{2N-1} f(x_p) \sin\left(\frac{2\pi k x_p}{L}\right), \quad k = 1, \ldots N-1 \quad (1.64)$$

It is obvious that if the continuous function contains harmonic components of wavelength less than $2\Delta x$, these cannot be included in the analysis in finite Fourier series. It has been proved that the corresponding coefficients of the analytical expansion in Fourier series are automatically embodied in the A_k, B_k coefficients of lower order harmonics. The introduction of noise in the numerical computation of A_k, B_k is defined as 'aliasing'[11].

20 ELEMENTS OF NUMERICAL ANALYSIS

The following relation holds between the A's (finite series) and the a's (infinite series):

$$A_k = a_k + \sum_{m=1}^{a} (a_{2Nm-k} + a_{2Nm+k}) \qquad (1.65)$$

If a_{2Nm+k} tends to zero for large m, then $A_k \to a_k$.

EXAMPLE 1.7

A digital water level recorder at a coastal location recorded over a period of 39 s the water surface levels at a rate of one measurement per second. The available short record is to be analysed in finite Fourier series. The recorded sample is as follows:

$f_i =$	0	1.7	2.3	1.3	−1	−2.5	−2	−1.1		−1.2	
	−1.3	0	1.5	2.4	2.4	1.2	−0.6	−2.5		−3.5	−3.1
	−1.6	0.5	2.6	3.3	2.5	1	−0.7	−0.8	0	1.5	2.0
	0.9	−0.5	−1.6	−2	−1.5	−0.4	0.9	2.0	1.7	0	

Equations 1.63 and 1.64 are programmed for the computation of A_K, B_K, as follows:

```
C     FINITE FOURIER SERIES
      DIMENSION F(100), A(100), B(100), X(100)
      READ(5,1) N,S
1     FORMAT(I4,F7.0)
      NN=2*N
      READ(5,11)(F(I), I=1,NN)
11    FORMAT (10F7.0)
      DO 2 I=1,NN
2     X(I)=S*(I−1)/NN
      N1=N+1
      DO 3 I=1,N1
      A(I)=0.
      DO 5 K=1,NN
5     A(I)=A(I)+F(K)*COS(6.2832/S*(I−1)*X(K))/N
3     CONTINUE
      N2=N−1
      DO 6 I=1,N2
      B(I)=0.
      DO 7 K=1,NN
7     B(I)=B(I)+F(K)*SIN(6.2832/S*(I)*X(K))/N
6     CONTINUE
      WRITE(6,8)(A(I), I=1,N1)
      WRITE(6,8)(B(I), I=1,N2)
8     FORMAT(10F10.4)
      STOP
      END
```

ELEMENTS OF NUMERICAL ANALYSIS 21

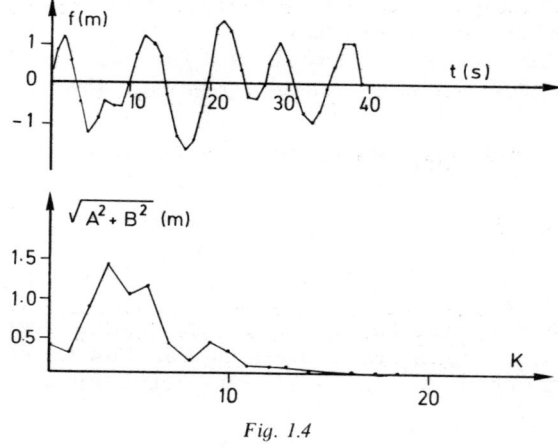

Fig. 1.4

Figure 1.4 contains the recorded signal and characteristic measure of the coefficients A_K, B_K ($K = 1 \ldots 20$) computed from the available 40 values.

1.7. METHOD OF FINITE DIFFERENCES IN THE SOLUTION OF DIFFERENTIAL EQUATIONS

Many problems in computational hydraulics require the numerical solution of differential equations. A classical way to approach those computational problems is through the method of finite differences.

The method consists of the approximate estimation of the values of one or more functions at characteristic locations (nodes) of the solution domain. This numerical estimation is achieved by discretisation of the domain through a one-, two- or three-dimensional grid, and the approximation of the differential equation by a difference equation.

The differential problem thus becomes an algebraic one and the computation of the field variable values at the nodes of the grid is a matter of the solution of algebraic equations, coupled or not.

The differential equations treated numerically are ordinary or partial ones. The most common elliptic, hyperbolic and parabolic differential equations will be investigated in detail later. The problems are posed as initial conditions, boundary conditions or mixed conditions.

There are many methods for the approximate substitution of the derivatives by algebraic differences. The fundamental procedure for the synthesis of finite differences numerical schemes derive from the known expansion in Taylor series.

If $f(x)$ has continuous derivatives up to order $n+1$, then its expansion in Taylor series around a point with abscissa $x=a$, is:

$$f(x) = f(a) + (x-a)f^{(1)}(a) + \frac{(x-a)^2}{2!} f^{(2)}(a) +$$

$$+ \cdots \frac{(x-a)^{n+1}}{(n+1)!} f^{(n+1)}(x_i) \qquad (1.66)$$

The last term is included due to the truncation of the series to the term of order n. The upper indices in parenthesis show the order of the derivative. For $x = x_0 + \Delta x$, the following relation holds:

$$f(x) = f(x_0) + \Delta x f^{(1)}(x_0) + \frac{\Delta x^2}{2!} f^{(2)}(x_0) + \cdots \qquad (1.67)$$

The use of the first three terms in Equation 1.67 results in the approximation of the first derivative by

$$\frac{df}{dx}(x_0) = \frac{f(x) - f(x_0)}{\Delta x} - \frac{\Delta x}{2} \frac{d^2 f}{dx^2}(x_1) \qquad (1.68)$$

The error in this forward difference is of order Δx.

For $x = x_0 - \Delta x$, Equation 1.67 becomes

$$f(x) = f(x_0) - \Delta x f^{(1)}(x_0) + \frac{\Delta x^2}{2!} f^{(2)}(x_0) + \cdots \qquad (1.69)$$

leading to the approximation of a first order derivative by a backward difference,

$$\frac{df}{dx}(x_0) = \frac{f(x_0) - f(x)}{\Delta x} + \frac{\Delta x^2}{2!} f^{(2)}(x_1) \qquad (1.70)$$

The subtraction of Equation 1.67 and 1.69 leads to the approximation of the first order derivative by a central difference

$$\frac{df}{dx}(x_0) = \frac{f(x_0 + \Delta x) - f(x_0 - \Delta x)}{2 \Delta x} + O(\Delta x^2) \qquad (1.71)$$

The error in this case is of order Δx^2.

The approximation of the second order derivative by finite differences is achieved by the addition of Equations 1.67 and 1.69.

$$\frac{d^2f}{dx^2}(x_0) = \frac{f(x_0 + \Delta x) - 2f(x_0) + f(x_0 - \Delta x)}{\Delta x^2} + O(\Delta x)^2 \quad (1.72)$$

Similarly for the 3rd order derivative it is found

$$\frac{d^3f}{dx^3}(x_0) = \frac{-f(x_0 - 2\Delta x) + 2f(x_0 - \Delta x) - 2f(x_0 + \Delta x) + f(x_0 + 2\Delta x)}{2\Delta x^3} +$$

$$+ O(\Delta x)^6 \quad (1.73)$$

The procedure is general and can supply derivatives by finite differences with approximations of any order. The accuracy depends on the location and the number of the field variable values involved in the approximation.

Algebraic equations relating the values of the field variable at the characteristic locations (grid points) can be formulated as an uncoupled or coupled system of simultaneous equations. Assume that the grid points are numbered in a certain sequence; then if each algebraic (difference) equation relates a new f value to preceding values of f, the computational scheme is called explicit. Otherwise, if the new value of f is related to both the preceding (known) and subsequent (unknown) f values the scheme is called implicit and requires the solution of an algebraic system of equations.

The approximation and substitution of differential operators by difference operators and the numerical integration on the discretised solution domain is not simple. Difficulties may arise whereby the computation of the field variable values on the grid points may lead to erroneous results or is impossible.

The quality of the finite differences numerical schemes can be checked on the basis of the following criteria:

(1) Consistency. A finite difference scheme is consistent if the finite differences operator has as limit the original differential operator when the discretisation step, Δx on the dimension x for example, tends to zero.

(2) Convergence. Convergence exists if the numerical solution tends to the analytical when $\Delta x \to 0$.

(3) Stability. The numerical scheme is stable if the error introduced by the numerical scheme remains bounded.

It should be mentioned here that the application of the electronic computer is not easy and does not always lead to correct results.

24 ELEMENTS OF NUMERICAL ANALYSIS

The questions which usually arise during the numerical solution of differential equations are:

(1) What is the optimal discretisation of the solution domain? The selection of Δx, Δy, Δz and the orientation of the coordinate axes has a decisive influence on the magnitude of the computational time and the numerical errors introduced.

There is an optimal step value minimising the total error (sum of the truncation and round off errors, the first proportional and the second inversely proportional to the step size).

(2) What is the proper finite difference scheme to be used? This selection also influences the computational time and the accuracy of the solution.

(3) What is the optimal way to tackle the initial and boundary conditions? There are two kinds of difficulties on the boundaries: a geometrical difficulty referring to the description of a curved boundary by a rectangular grid and the error involved in expressing the function by its value at neighbouring points.

(4) What is the proper computer to be used? The arithmetic operations to be performed requires a certain storage capacity and speed. Also the numerical errors arising depend on the precision implicit in each machine.

An application of the finite difference method in the solution of differential equations is included in the following example.

EXAMPLE 1.8

To compute the time taken to empty a cylindrical tank. The initial depth is $h_0 = 5$ m.

The ratio of the horizontal section of the tank, S_1, to the section, S_2, at the opening on the bottom of the tank is 987:1. It is assumed that the Torricelli equation holds for the velocity of the water at the exit and the discharge coefficient is $\mu = 1$.

The discharge through the bottom opening $Q = US_2 = \sqrt{(2gh)}S_2$ is equal to the rate of variation of the water depth in the tank, due to the mass conservation principle. The emptying process, therefore, can be described quantitatively by the equation,

$$\sqrt{(2gh)}S_2 = -S_1 \frac{dh}{dt} \qquad (1.74)$$

which can be written as

$$\frac{dh}{dt} = -\sqrt{(2g)}\frac{S_2}{S_1}\sqrt{h} = -a\sqrt{h} = -0.004488\sqrt{h} \qquad (1.75)$$

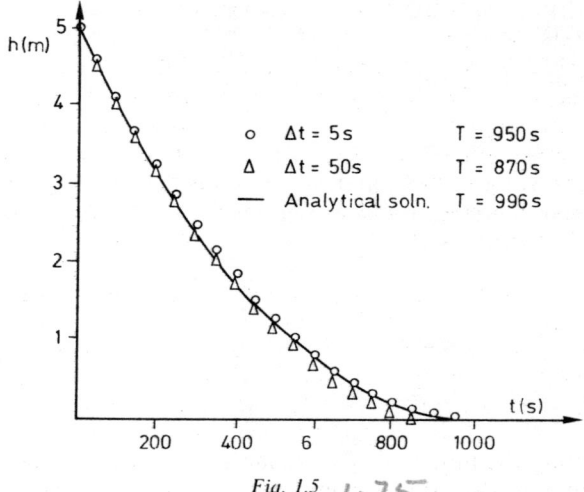

Fig. 1.5 1.75

The analytical solution of Equation 2.75 gives for the $h(t)$ function,

$$h = \left(\sqrt{h_0} - \frac{a}{2}t\right)^2 \quad (1.76)$$

and an emptying time T (for $h = 0$) = 996 s.

The numerical integration by finite differences requires the approximation of the differential equation. If the upper index n is the time index ($t_n = n\Delta t$, where Δt is the step used in the time dimension) the application of Equation 1.68, the forward difference equation, to Equation 1.75 leads to,

$$(h^{n+1} - h^n)/\Delta t = -a\sqrt{h^n} \quad (1.77)$$

This can be solved explicitly for h^{n+1} in terms of h^n.

The initial condition $h_0 = 5$ m permits the explicit computation of $h^1, h^2 \ldots$. The repetitive application of Equation 1.77 can be programmed as follows:

```
C      NUMERICAL SOLUTION OF ORDINARY DIF EQUATION
       DIMENSION H(500)
       READ (5,1) DT, A, HH
1      FORMAT(3F7.0)
       H(1)=HH
       DO 2 I=2,500
       H(I)=H(I−1)−DT*A*SQRT(H(I−1))
```

```
      IF(H(I).LE..001) GO TO 4
 2    CONTINUE
 4    WRITE(6,3)(H(N),N=1,I)
 3    FORMAT(10F10.3)
      STOP
      END
```

The solutions for $\Delta t = 5$ s and $\Delta t = 50$ s are depicted in Fig. 1.5 and compared to the analytical solution. The influence of the time step on the emptying time is evident.

2

Numerical solution of partial differential equations common in hydraulics

2.1. FORM AND OCCURRENCE OF SOME PARTIAL DIFFERENTIAL EQUATIONS

Most mathematical models describing the motion and equilibrium of fluids are composed of differential equations deriving from the application of the fundamental principles of mass continuity and energy conservation, together with the appropriate initial and boundary conditions. In one dimension these are ordinary differential equations, in two or more dimensions, partial differential equations.

This chapter will be devoted to the numerical solution of some typical PDEs frequently appearing in hydraulic mathematical models. The equations to be treated can be classified in two main groups: those describing a situation of equilibrium or steady flow and those describing transient phenomena. The former do not contain the time variable, t, and are formulated as boundary value problems, the latter contain the variable t and are usually formulated as mixed initial and boundary value problems.

Usually these equations take the form of linear or non-linear, homogeneous or non-homogeneous second order partial differential equations. The form of a linear homogeneous 2nd order PDE is:

$$a \frac{\partial^2 f}{\partial x^2} + b \frac{\partial^2 f}{\partial y \partial x} + c \frac{\partial^2 f}{\partial y^2} = 0 \qquad (2.1)$$

From the theory of PDEs it is known that Equation 2.1 can take three, distinct forms according to the value of the discriminant, $b^2 - 4ac$:

(1) If $b^2 - 4ac > 0$, Equation 3.1 is called a hyperbolic equation and can take the canonical form:

$$\frac{\partial^2 f}{\partial x^2} - c^2 \frac{\partial^2 f}{\partial y^2} = 0 \qquad (2.2)$$

Equation 2.2 is known as the 'wave equation' which describes the propagation of an initial signal $f_0 = f(0, y)$, undeformed, in the positive and negative y directions with a speed of propagation equal to c.

(2) If $b^2 - 4ac = 0$, Equation 2.1 is called a parabolic equation, and can take the canonical form

$$\frac{\partial f}{\partial x} = c \frac{\partial^2 f}{\partial y^2} \qquad (2.3)$$

Equation 2.3 is known as the 'heat equation' which describes the diffusion and decay of an initial signal $f_0 = f(0, y)$ at a rate related to the value of the diffusion coefficient c.

(3) If $b^2 - 4ac < 0$, Equation 2.1 is called an elliptic equation and reduces to the canonical form

$$\frac{\partial^2 f}{\partial x^2} + \frac{\partial^2 f}{\partial y^2} = 0 \qquad (2.4)$$

Equation 2.4 is known as the 'Laplace equation' which describes the distribution under equilibrium of a function $f(x, y)$ over a two-dimensional field where the values of f or its derivatives are known on the boundaries.

The above three PDEs describe, from a physical point of view, three different physical processes and according to the nature of the physical phenomenon to be modelled they appear separately or simultaneously in various combinations.

Elliptic equations describe an equilibrium or a steady flow. The f values represent potential values and derive from the principle of the minimisation of the energy of the system. They are usually met in hydraulic problems of steady flow of an ideal fluid and the flow in porous media.

Parabolic equations describe a diffusion or decay process of a certain physical magnitude, such as momentum, temperature or the concentration of a substance, etc. They are usually met in problems of unsteady flow in porous media and of diffusion–disperson phenomena in open channels.

Finally, hyperbolic equations describe the propagation of a signal in several directions and are met in most of the phenomena of non-steady flow in open and closed conduits and in marine hydrodynamics (waves and tides).

2.2. NUMERICAL SOLUTION OF PARABOLIC EQUATIONS

The simplest form of the parabolic equation in two-dimensions, x, t, is:

$$\frac{\partial f}{\partial t} = c \frac{\partial^2 f}{\partial x^2} \qquad (2.5)$$

It has been mentioned that the rate of dissipation of the scalar magnitude f along x with time t depends on the value of the diffusion coefficient, c. The boundary conditions can have the form of two prescribed values or derivatives of f at the ends of the field (two such values are required as the equation is of 2nd order with respect to x). The initial conditions usually have the form of known f values at time $t = 0$, $f_0 = f(x, 0)$.

The numerical solution of Equation 2.5 requires the application of the method of finite differences and can be achieved through several explicit and implicit finite difference schemes. The one most commonly used derives from an approximation of the time derivative by a forward difference and the spatial second derivative by a central difference. If the time is indicated by the upper index n ($t_n = n\Delta t$, where Δt is the time step used for the discretisation of time) and the position along the x axis by the lower index i ($x_i = i\Delta x$, where Δx is the space step), Equation 2.5 can be approximated by,

$$\frac{f_i^{n+1} - f_i^n}{\Delta t} = c \frac{f_{i+1}^n - 2f_i^n + f_{i-1}^n}{\Delta x^2} \qquad (2.6)$$

The following example comprises a computer program for the integration of the parabolic equation by means of the explicit scheme shown and points out the problems that may arise.

EXAMPLE 2.1

The space between two parallel horizontal plates is filled with liquid of kinematic viscosity coefficient equal to $v = 1 \text{ cm}^2/\text{s}$. The distance between the plates is 1 cm. The upper plate is linearly accelerated from a zero initial velocity to 10 cm/s in 0.1 s. The change with time of the velocity profile of the liquid is to be computed.

The conservation of momentum in the horizontal direction x, according to the notation of Fig. 2.1, taking into consideration that

30 NUMERICAL SOLUTION OF PARTIAL DIFFERENTIAL EQUATIONS

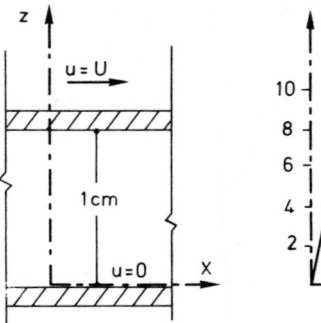
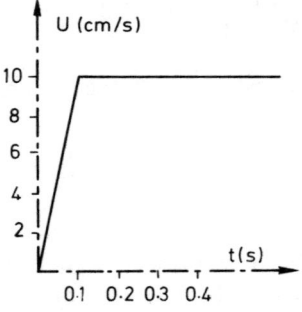

Fig. 2.1

there is no variation in the x, y directions, i.e. one-dimensional case, is expressed by the parabolic equation w.r.t. the velocity function $u(z, t)$,

$$\frac{\partial u}{\partial t} = v \frac{\partial^2 u}{\partial z^2} \tag{2.7}$$

The flow domain is bounded between $z = 0$ cm and $z = 1$ cm. The initial condition has the form $u(z, 0) = 0$ cm/s. The boundary conditions are $u|_{z=0} = 0$, $u|_{z=1} = U$ cm/s, where U is the upper plate velocity. The change in U with time is shown in Fig. 2.1.

The numerical integration is performed by means of the explicit finite differences scheme described in Equation 2.6, with $\Delta z = 0.1$ cm and $\Delta t = 0.001$ s or 0.01 s. The space discretisation implies that the velocity is computed at 11 equidistant points between the plates along a straight line normal to the plates. The velocity is given by the boundary conditions at points $i = 1$ and $i = 11$.

The FORTRAN program for the integration can have the form:

```
C     SOLUTION OF PARABOLIC EQN. EXPLICIT SCHEME
      DIMENSION U(100), UO(100)
      READ(5,1) DX, DT, IMAX
1     FORMAT(2F7.0, I4)
      IMAX1 = IMAX - 1
      DO 2 I = 1, IMAX
      U(I) = 0.
2     UO(I) = 0.
      T = 0.
      N = 0
100   T = T + DT
      N = N + 1
      IF(T.LT..1) GO TO 22
      U(IMAX) = 10.
      GO TO 23
```

```
22  U(IMAX)=T*100.
23  CONTINUE
    DO 3 I=2, IMAX1
 3  U(I)=UO(I)+DT*1./DX**2*(UO(I+1)+UO(I-1)
    -2.*UO(I))
    DO 4 I=1,IMAX
 4  UO(I)=U(I)
    WRITE(6,5)(UO(I), I=1,IMAX)
 5  FORMAT(10F10.4)
    IF(N.LT.1000) GO TO 100
    STOP
    END
```

Variables description:
U(I), UO(I) = recent and past velocity values
DX   = space step
DT   = time step
IMAX   = max value of space index I
T   = time
N   = time index

Data values:
DX = 0.1, DT = 0.001, and 0.01, IMAX = 11

The change with time of the velocity profile for DT = 0.001 is shown in Fig. 2.2.

By comparing the change of velocity values at $z = 0.9$ cm, for DT = 0.001 and DT = 0.01, corresponding to the parameter

$$\lambda = v \frac{\Delta t}{\Delta z^2} \tag{2.8}$$

equal to 0.1 and 1.0, respectively, it is evident that the solution is unstable in the second case. Numerical experiments and theoretical analysis show that the explicit scheme is stable for $\lambda \leqslant 0.5$.

Apart from the explicit scheme described in Example 3.1, many other finite difference schemes have been devised and tested by various researchers. The Duffort–Frankel scheme, for example, approximates the second derivative of Equation 2.5 by the difference,

$$\frac{\partial^2 f}{\partial x^2} \approx \frac{f_{i+1}^n + f_{i-1}^n - f_i^{n+1} - f_i^{n-1}}{\Delta x^2} \tag{2.9}$$

and the Cheng–Allen scheme by

$$\frac{\partial^2 f}{\partial x^2} \approx \frac{f_{i+1}^n + f_{i-1}^n - 2f_i^{n-1}}{\Delta x^2} \tag{2.10}$$

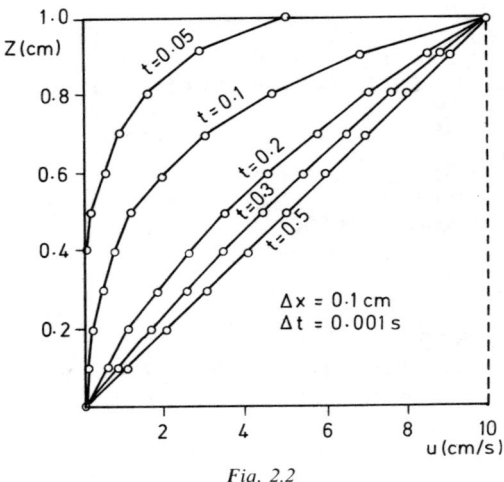

Fig. 2.2

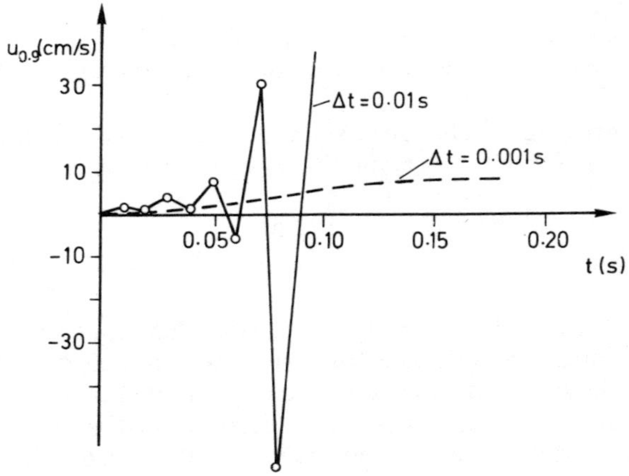

Fig. 2.3

A slight problem arises during the computation of f for $n=1$ ($t=\Delta t$). Then a knowledge of f, for $t=0$ and $t=-\Delta t$, is required. An explicit scheme of the form of Equation 2.6 can be initially used, followed by either Equations 2.9 or 2.10.

Contrary to the above explicit schemes an implicit one, known as the Crank–Nicolson method, approximates the second derivative by the difference,

where
$$\frac{\partial^2 f}{\partial x^2} \approx \frac{f_i^{n+\frac{1}{2}} + f_{i-1}^{n+\frac{1}{2}} - 2f_i^{n+\frac{1}{2}}}{\Delta x^2}$$

$$f^{n+\frac{1}{2}} = \frac{f^n + f^{n+1}}{2} \tag{2.11}$$

It is not characterised by any stability limitation dependent on λ values, but its application leads to the solution of a tri-diagonal system of algebraic equations containing the f_i^{n+1} values on the whole solution domain ($i = 1 \ldots i_{max}$); the system can be solved according to the forementioned methods. The following example contains the computer program and the results of the application of the Crank–Nicolson method in the previous problem.

EXAMPLE 2.2

Example 2.1 is to be repeated. The integration is to be performed by means of the Crank–Nicolson finite difference method.

```
C     SOLUTION OF PARABOLIC EQN. CRANK NICOLSON
C     SCHEME
      DIMENSION U(100), UO(100)
      READ(5,1) DX,DT,IMAX
   1  FORMAT(2F7.0,I4)
      DO 2 I=1,IMAX
      U(I)=0.
   2  UO(I)=0.
      IMAX1=IMAX-1
      N=0
      T=0.
      B=1./(1./DT+1./DX**2)
 100  T=T+DT
      N=N+1
      IF(T.LT..1) GO TO 23
      U(1)=10.
      GO TO 22
  23  U(1)=T*100.
  22  CONTINUE
      ITER=0
 101  DIFMX=0.
      ITER=ITER+1
      IF(ITER.GT.100) STOP 1
      DO 3 I=2, IMAX1
      TEMP=U(I)
      U(I)=UO(I)*B/DT+1.*B/2./DX**2*(U(I+1)+
     U(I-1)+UO(I+1)+UO(I-1)-2.*UO(I))
      DIF=ABS(TEMP-U(I))
      IF(DIF.GT.DIFMX) DIFMX=DIF
```

```
3   CONTINUE
    IF (DIFMX.GT..00001) GO TO 101
    DO 4 I=1, IMAX
4   UO(I)=U(I)
    WRITE(6,5)(UO(I),I=1,IMAX)
5   FORMAT(10F10.4)
    IF(N.LT.1000) GO TO 100
    STOP
    END
```

Description of additional variables:
ITER = number of iterations in the solution of the algebraic system.
DIFMX = maximum difference of U values between two iterations; the convergence criterion is $DIFMX < 10^{-5}$.

As can be seen from Fig. 2.4, representing graphically the change in velocity at level $z = 0.5$ cm, there is no major difference between the two methods.

A variety of other numerical schemes exist for the integration of the parabolic equation and selection of the method to be used for each special case can be an optimisation problem.

The parabolic equation in two dimensions (x, y, t),

$$\frac{\partial f}{\partial t} = c \frac{\partial^2 f}{\partial x^2} + c \frac{\partial^2 f}{\partial y^2} \tag{2.12}$$

is treated in the same way. The application of the explicit scheme presented permits the direct solution of the deriving uncoupled algebraic equations for the computation of f^{n+1} values from f^n. If the field discretisation is done by means of a grid of orthogonal meshes, with sides Δx, Δy ($x_i = i\Delta x$, $y_j = j\Delta y$, $i = 1, i_{max}, j = 1, j_{max}, t_n = n\Delta t$), the deriving difference equation approximating Equation 2.12, in the case of $\Delta x = \Delta y$, is,

$$\frac{f_{i,j}^{n+1} - f_{i,j}^n}{\Delta t} = \frac{c}{\Delta x^2} (f_{i,j+1}^n + f_{i,j-1}^n - 4f_{i,j}^n + f_{i+1,j}^n + f_{i-1,j}^n) \tag{2.13}$$

The application of the Crank–Nicolson or any other implicit scheme, permitting the increase of the Δt step beyond the limit posed by Equation 2.8, leads to a systems of simultaneous algebraic equations. The number of unknowns is equal to the number of the grid nodal points. For large 2-D or 3-D solution domains this number becomes so large that, even if it is not prohibitive for contemporary computers, it makes the numerical solution computationally uneconomical. This problem is solved by the introduction of the ADI method (Alternating Directions Implicit). The technique, as its name

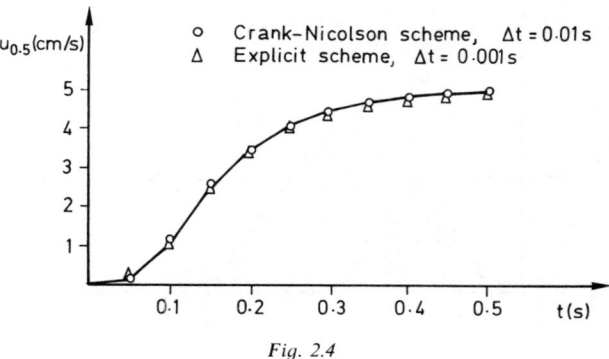

Fig. 2.4

suggests, consists of alternately applying implicit and explicit numerical schemes along the x and y dimensions. This results in the need to solve, alternately, a j_{max} number of systems of equations in i_{max} unknowns at one time step and an i_{max} number of systems of equations in j_{max} unknowns at the next time step. If the Crank–Nicolson scheme is used interchangeably with an explicit method, the ADI procedure can be described by the following equations:

$$\frac{f^{n+1}-f^n}{\Delta t}=c\frac{\partial^2 f^{n+\frac{1}{2}}}{\partial x^2}+c\frac{\partial^2 f^n}{\partial y^2} \qquad (2.14)$$

and subsequently

$$\frac{f^{n+2}-f^{n+1}}{\Delta t}=c\frac{\partial^2 f^{n+1}}{\partial x^2}+c\frac{\partial^2 f^{n+3/2}}{\partial y^2} \qquad (2.15(a))$$

where

$$f^{n+\frac{1}{2}}=(f^n+f^{n+1})/2 \quad f^{n+3/2}=(f^{n+2}+f^{n+1})/2 \qquad (2.15(b))$$

2.3. THE HYPERBOLIC EQUATION AND THE METHOD OF CHARACTERISTICS

The canonical form of the homogeneous hyperbolic equation in one dimension, is

$$\frac{\partial^2 f}{\partial t^2}-c^2\frac{\partial^2 f}{\partial x^2}=0 \qquad (2.16)$$

With the introduction of the auxiliary variables $\xi=x+ct$, $\eta=x-ct$

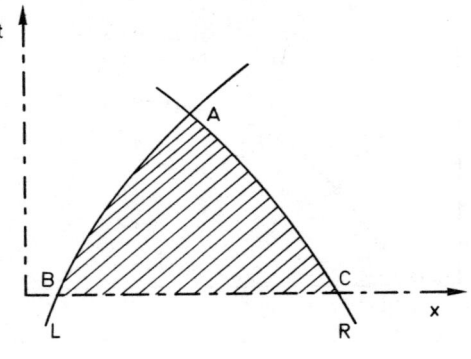

Fig. 2.5

Equation 2.16 becomes,

$$\frac{\partial^2 f}{\partial \xi \partial n} = 0 \qquad (2.17)$$

showing that f has the general form

$$f = \varphi(x + ct) + \psi(x - ct) \qquad (2.18)$$

From a geometrical point of view Equation 2.18 reveals that hyperbolic equations have the following property: on certain families of curves or surfaces in the solution domain, called characteristic curves (or surfaces), the solution of Equation 2.16 varies according to a specific law. In the case of the 1-D wave equation, Equation 2.16, the families of curves are lines described by the equations

$$\frac{dx}{dt} = \pm c \qquad (2.19)$$

and f is constant along these characteristic lines.

The above property, i.e. the existence of characteristic curves, is very useful for the analytical or numerical solution of hyperbolic equations. The determination of the characteristic curves and the subsequent solution of the equation either on these characteristic lines or on a known rectangular grid (but taking into consideration the property of characteristics) is the best way to avoid errors. If, for example, point A of Fig. 2.5 is crossed by the characteristic lines AR and AL, the field variable value f_A depends on the f values within the domain ABC, and any numerical integration scheme not complying with this geometric condition must result in numerical instability (suppressed only by numerical dissipation, artificially introduced, as will be shown).

The 'first order hyperbolic equation' (known as the 'colour equation') will be analysed in detail, as a typical hyperbolic equation,

$$\frac{\partial f}{\partial t} + c \frac{\partial f}{\partial x} = 0 \tag{2.20}$$

If c is replaced by

$$c = dx/dt \tag{2.21}$$

having as solution $x = x_0 + ct$, Equation 2.20 takes the form

$$\frac{\partial f}{\partial t} + \frac{dx}{dt} \frac{\partial f}{\partial x} = \frac{df}{dt} = 0 \tag{2.22}$$

showing that f is constant (has a zero derivative) along each of the characteristic lines $x = x_0 + ct$. The colour equation, Equation 2.20, forms a part of the wave equation as it describes the propagation of a signal along one direction in space.

If the initial condition has the form $f(x, 0) = f_0(x)$, the general solution of Equation 2.20 is

$$f = f_0(\xi) \quad \text{where} \quad \xi = x + ct \tag{2.23}$$

The numerical solution of Equation 2.20 can be easily checked and controlled by comparison with the analytical solution, Equation 2.33. The control can focus on the speed of propagation (phase error) or the change in the initial signal, f_0, due to numerical errors. The deformation usually appearing, is due to the dissipation of the f_0 values and the spatial spreading along the x dimension (numerical diffusion) as well as the appearance at the beginning or at the end of the signal of small oscillations due to the difference in the speed of propagation of the various harmonic components included in f_0 (numerical dispersion). The change in the wave shape due to the numerical solution is qualitatively illustrated in Fig. 2.6.

The fact that the family of characteristics for the colour equation is given by Equation 2.2 implies that the f values at point A depend on the f values within the shaded area ABC of the x–t plane, Fig. 2.7. This basic property, revealing itself only through the study of the characteristic curves, plays a very important role in the synthesis of numerical methods for the integration of Equation 2.20.

38 NUMERICAL SOLUTION OF PARTIAL DIFFERENTIAL EQUATIONS

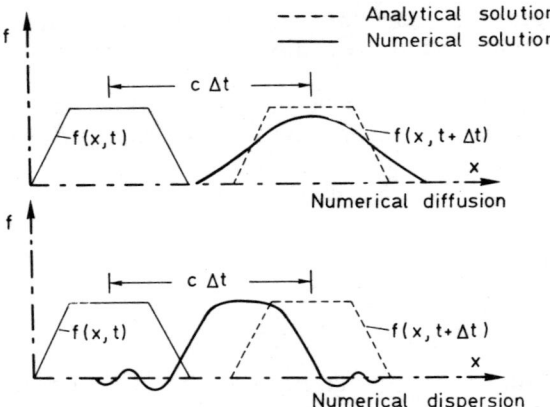

Fig. 2.6

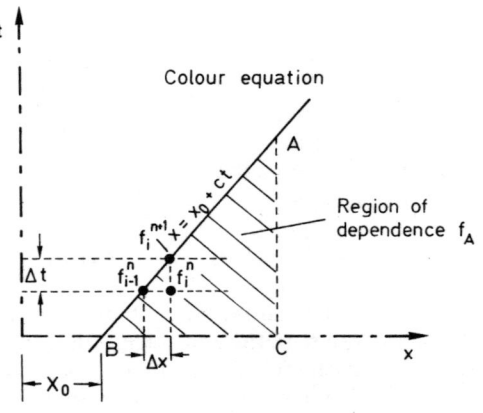

Fig. 2.7

The approximation of the spatial derivative by a forward or central difference

$$\frac{\partial f}{\partial x} \approx \frac{f_{i+1}^n - f_i^n}{\Delta x} \quad \text{or} \quad \frac{\partial f}{\partial x} \approx \frac{f_{i+1}^n - f_{i-1}^n}{2\Delta x} \qquad (2.24)$$

correlating the f_i^{n+1} value to f values lying to the right w.r.t. f_i^{n+1} in the x–t space, results in numerical instability. To the contrary, the approximation of $\partial f/\partial x$ by backward difference leads to a stable numerical scheme

$$\frac{f_i^{n+1}-f_i^n}{\Delta t}+c\frac{f_i^n-f_{i-1}^n}{\Delta x}=0 \tag{2.25}$$

It can be seen from Fig. 2.7, graphically deploying the property of characteristics of Equation 2.20, that in order to correlate f_i^{n+1} to f_i^n, f_{i-1}^n values inside the shaded area, the term $c\Delta t/\Delta x$ must satisfy the inequality

$$\lambda=\frac{c\Delta t}{\Delta x}\leqslant 1 \tag{2.26}$$

where c is the speed of propagation of the signal.

A theoretical investigation of numerical stability leads to the same conclusion, known as the Courant criterion, for the numerical integration of hyperbolic equations. In Example 2.3, the integration of the colour equation by an explicit finite differences scheme is given.

EXAMPLE 2.3

Describe the change with time of a function f satisfying the colour equation 2.20, with $c=2$. The initial form of f is trapezoidal; $f(0.0)=0$, $f(1.0)=1$, $f(2.0)=1$, $f(3.0)=0$.

The integration is performed numerically by means of an explicit backward differences scheme with $\Delta x=1$, $\Delta t=0.5$ and $\Delta t=0.25$ (corresponding to λ values, $\lambda=1$, $\lambda=1/2$, respectively).

```
C      SOLUTION OF HYPERBOLIC EQN EXPLICIT SCHEME
       DIMENSION F(100), FO(100)
       READ(5,1) DT,DX,C,IMAX
 1     FORMAT(3F7.0, I4)
       DO 7 I=1,IMAX
 7     FO(I)=0.
       READ(5,6)(FO(I),I=1,4)
 6     FORMAT(4F7.0)
       DO 8 I=1,IMAX
 8     F(I)=FO(I)
       N=0
       T=0.
 100   T=T+DT
       N=N+1
       DO 9 I=2,IMAX
 9     F(I)=FO(I)-C*DT/DX*(FO(I)-FO(I-1))
       DO 10 I=1,IMAX
 10    FO(I)=F(I)
       WRITE(6,11)(FO(I),I=1,IMAX)
```

```
11  FORMAT(10F10.4)
    IF(N.LT.100) GO TO 100
    STOP
    END
```

From a plot of the solution for $\Delta t = 0.5$, $\Delta t = 0.25$, shown in Fig. 2.8, it is concluded that:
  (1) The scheme is stable.
  (2) It conserves the symmetric form of the signal.
  (3) It is characterised by a tendency to diffuse the f_0 values.
  (4) The inherent numerical diffusion is inversely proportional to the value of λ. The numerical solution coincides with the analytical solution for $\lambda = 1$.
  (5) No dispersion is observed.

An unstable method can be transformed into a stable one through the introduction of numerical diffusion. A method has been developed on this principle, referred to as the Lax–Wendroff method.

The use of central differences for the space derivative w.r.t. x and the approximation of the time derivative by

$$\frac{\partial f}{\partial t} = \frac{f_i^{n+1} - \tfrac{1}{2}(f_{i+1}^n + f_{i-1}^n)}{\Delta t} \tag{2.27}$$

leads to a numerical scheme stable for $\lambda < 1$.

$$\frac{f_i^{n+1} - \tfrac{1}{2}(f_{i-1}^n + f_{i+1}^n)}{\Delta t} + c\left(\frac{f_{i+1}^n - f_{i-1}^n}{2\Delta x}\right) = 0 \tag{2.28}$$

Equation 2.28 can be written as,

$$\frac{f_i^{n+1} - f_i^n}{\Delta t} + c\left(\frac{f_{i+1}^n - f_{i-1}^n}{2\Delta x}\right) - \frac{\Delta x^2}{2\Delta t}\left(\frac{f_{i+1}^n - 2f_i^n + f_{i-1}^n}{\Delta x^2}\right) = 0 \tag{2.29}$$

showing that the above numerical method, known as the Lax–Wendroff method, artificially introduces a second order term $c\partial^2 f/\partial x^2$ to the colour equation where the numerical diffusion coefficient c is proportional to $\Delta x^2/2\Delta t$.

The expansion in Taylor series with respect to the time variable is

$$f_i^{n+1} = f_i^n + \Delta t \left.\frac{\partial f}{\partial t}\right|_i^n + \frac{\Delta t^2}{2}\left.\frac{\partial^2 f}{\partial t^2}\right|_i^n + O(\Delta t^3) \tag{2.30}$$

Expressing $\partial f/\partial t$, $\partial^2 f/\partial t^2$ by means of the spatial derivatives of f, with

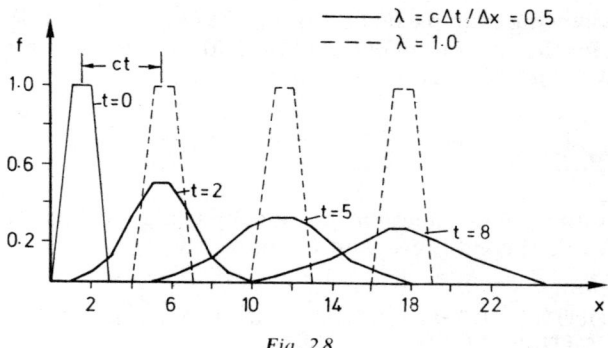

Fig. 2.8

respect to x, through the use of the colour Equation 2.20, gives

$$\left.\frac{\partial f}{\partial t}\right|_j^n = -c\left.\frac{\partial f}{\partial x}\right|_j^n = -c\frac{f_{i+1}^n - f_{i-1}^n}{2\Delta x} \tag{2.31}$$

and

$$\left.\frac{\partial^2 f}{\partial t^2}\right|_j^n = \left.\frac{\partial}{\partial t}\left(-c\frac{\partial f}{\partial x}\right)\right|_j^n = -c\left.\frac{\partial}{\partial x}\left(\frac{\partial f}{\partial t}\right)\right|_j^n = +c^2\left.\frac{\partial^2 f}{\partial x^2}\right|_j^n =$$

$$= c^2 \frac{f_{i+1}^n + f_{i-1}^n - 2f_i^n}{\Delta x^2} \tag{2.32}$$

The subsequent substitution of Equations 2.31 and 2.32 in Equation 2.30 leads to the Lax–Wendroff equation,

$$f_i^{n+1} = f_i^n - \frac{\lambda}{2}(f_{i+1}^n - f_{i-1}^n) + \frac{\lambda^2}{2}(f_{i+1}^n - 2f_i^n + f_{i-1}^n) \tag{2.33}$$

The artificially introduced numerical diffusion is proportional to $c^2 \Delta t/2$. The stability control results in the same Courant criterion, $\lambda < 1$, as before.

The same behaviour characterises the 'leap frog' method approximating both the time and space derivatives, by central differences

$$\frac{f_i^{n+1} - f_i^{n-1}}{2\Delta t} + c\frac{f_{i+1}^n - f_{i-1}^n}{2\Delta x} = 0 \tag{2.34}$$

The following example demonstrates the use of the Lax–Wendroff method for the integration of Equation 2.20 with the same initial and boundary conditions as in Example 2.3.

EXAMPLE 2.4

To solve the colour equation numerically using the Lax–Wendroff method. Initial conditions as in Example 2.3.

```
C     SOLUTION OF HYPERBOLIC EQN. LAX WENDROFF SCHEME
      DIMENSION F(100), FO(100)
      READ(5,5) DT,DX,C,IMAX
 5    FORMAT(3F7.0,I4)
      IMAX1=IMAX−1
      DO 6 I=1,IMAX.
 6    FO(I)=0.
      READ(5,7)(FO(I),I=1,4)
 7    FORMAT(4F7.0)
      DO 8 I=1,IMAX
 8    F(I)=FO(I)
      EL=C*DT/DX
      T=0.
      N=0
100   T=T+DT
      N=N+1
      DO 10 I=2,IMAX1
 10   F(I)=FO(I)−EL/2.*(FO(I+1)−FO(I−1))+
     1EL**2/2.*(FO(I+1)+FO(I−1)−2.*FO(I))
      DO 11 I=1,IMAX
 11   FO(I)=F(I)
      WRITE(6,12)(FO(I),I=1,IMAX)
      IF(N.LT.100) GO TO 100
 12   FORMAT(10F10.4)
      STOP
      END
```

The use of small $\Delta t = 1/4$ ($\lambda = 1/2$) keeps numerical diffusion to a low level and at the same time reveals the numerical dispersion of centered differences in hyperbolic equations.

The various harmonics included in f_0 propagate at different speeds and the phase error, shown in the form of a trace in Fig. 2.9, graphically represents the solution (the numerical solution lagging the analytical solution).

Numerical dispersion can be suppressed using a method due to Fromm. The method consists of the superposition of two finite difference methods with phase errors of different sign (the one leading

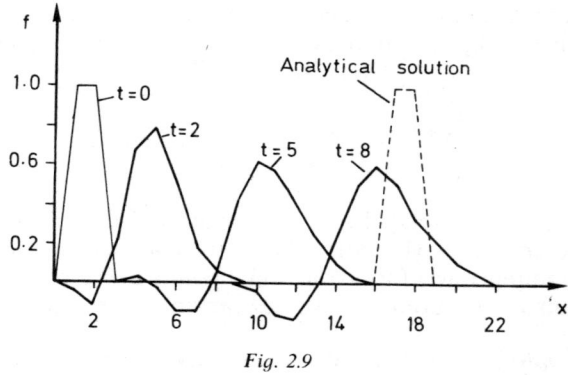

Fig. 2.9

and the other lagging the signal). Without details of derivation the Fromm method approximates the colour equation, by

$$f_i^{n+1} = f_i^n - \frac{c\Delta t}{4\Delta x}(f_{i+1}^n - f_{i-1}^n + f_i^n - f_{i-2}^n) + \frac{c^2 \Delta t^2}{4\Delta x^2}(f_{i+1}^n - 2f_i^n + f_{i-1}^n) +$$

$$+ \frac{c^2 \Delta t^2 - 2c\Delta t \Delta x}{4\Delta x^2}(f_{i-2}^n - 2f_{i-1}^n + f_i^n) \tag{2.35}$$

2.4. ELLIPTIC EQUATIONS

The typical form of the elliptic equation in two dimensions

$$\nabla^2 f = \frac{\partial^2 f}{\partial x^2} + \frac{\partial^2 f}{\partial y^2} = \varphi(x, y) \tag{2.36}$$

is known as Poisson's equation. The problem is well posed as a boundary value problem.

The three most common boundary conditions are:
(1) Dirichlet type boundary conditions, where the f values are known on the boundaries of the solution domain..
(2) Newmann type boundary conditions, when the gradient of f normal to the boundaries are known together with a value of f.
(3) Mixed boundary conditions, when some f values and some gradient values are known around the solution domain.

The most common procedure for the numerical solution of elliptic equations is the approximation of the second order derivatives by central finite differences. The resulting method is an implicit one. For equal space discretisation steps, $\Delta x = \Delta y$, and denoting the abscissae and ordinates by the indices i, j ($x_i = ix$, $y_j = jy$), Equation 2.36 becomes

$$\frac{f_{i+1,j} + f_{i-1,j} + f_{i,j+1} + f_{i,j-1} - 4f_{i,j}}{\Delta x^2} = \varphi_{i,j} \qquad (2.37)$$

As the $f_{i,j}$ value is related to f values around the i, j location, the difference equations (Equation 2.37) are coupled and have to be solved simultaneously for all the nodal points of the grid established inside the solution domain (i values ranging from 1 to i_{max} and $j = 1 \ldots j_{max}$).

The resulting system of algebraic equations has a tri-diagonal coefficient matrix. Some of the elements may themselves be matrices. The most common procedures for its solution are:

(1) Solution by Gauss elimination.
(2) Solution by successive over relaxation (SOR). The residual $R_{i,j}$,

$$R_{i,j}^{(v)} = \frac{f_{i+1,j}^{(v)} + f_{i-1,j}^{(v)} + f_{i,j+1}^{(v)} + f_{i,j-1}^{(v)} - 4f_{i,j}^{(v)}}{\Delta x^2} - \varphi_{i,j} \qquad (2.38)$$

where the upper index (v) shows the order of iteration, is calculated on each grid point and the $f^{(v)}$ values are corrected to new ones $f^{(v+1)}$, where

$$f_{i,j}^{(v+1)} = f_{i,j}^{(v)} + \frac{\omega}{4} R_{i,j}^{(v)} \qquad (2.39)$$

and ω is the relaxation coefficient with an optimal value, $\omega = 1.8$ to 2.0, ensuring quick convergence.

(3) Gauss–Seidel iteration method. According to this method the grid points are numbered to run sequentially and the computation of $f_{i,j}$ values, by means of Equation 2.37, propagates along the x, y directions (i.e. continuously increasing i, j values) making use of the most recent available f values.

$$f_{i,j}^{(v+1)} = \frac{f_{i-1,j}^{(v+1)} + f_{i,j-1}^{(v+1)} + f_{i,j+1}^{(v)} + f_{i+1,j}^{(v)}}{4} - \varphi_{i,j} \frac{\Delta x^2}{4} \qquad (2.40)$$

(4) Carré method, a special form of the successive over relaxation method (SOR). According to this method the new f values are given by

NUMERICAL SOLUTION OF PARTIAL DIFFERENTIAL EQUATIONS 45

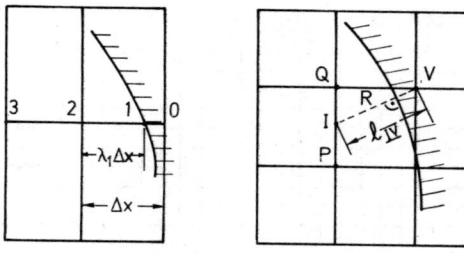

Fig. 2.10

$$f_{i,j}^{(v+1)} = (1-\omega)f_{i,j}^{(v)} + \omega\left(\frac{f_{i-1,j}^{(v)} + f_{i+1,j}^{(v)} + f_{i,j-1}^{(v)} + f_{i,j+1}^{(v)}}{4} - \varphi_{i,j}\frac{\Delta x^2}{4}\right) \quad (2.41)$$

with an appropriate estimation of the relaxation coefficient ω.

A general problem, which arises due to the morphological complexity of the solution domains for all types of parabolic, hyperbolic and elliptic equations, is the approximation of curved boundaries and the adaptation of the difference equations on those boundaries.

This is a very frequent situation in hydraulics where the geophysical flow domains have a complex geometry.

The termination, for example, of the flow field at point 1 of Fig. 2.10, instead of point 0 of the grid, requires either the artificial extension of the field to 0 or the proper adaptation of the finite differences formulae as described below.

Through the expansion in Taylor series of the field variable at points 1 and 3 we can approximate the derivatives at 2 as follows:

$$\left.\frac{\partial f}{\partial x}\right|_2 = \frac{1}{\Delta x}\left(\frac{f_1}{(\lambda_1+1)\lambda_1} - \frac{\lambda_1 f_3}{1+\lambda_1} - \frac{(1-\lambda_1)f_2}{\lambda_1}\right) \quad (2.42)$$

$$\left.\frac{\partial^2 f}{\partial x^2}\right|_2 = \frac{1}{\Delta x^2}\left(\frac{2f_1}{\lambda_1(\lambda_1+1)} + \frac{2f_3}{1+\lambda_1} - \frac{2f_2}{\lambda_1}\right) \quad (2.43)$$

The f_2 value is expressed through Equations 2.42 and 2.43 when the f_1 value is given. When the boundary condition refers to the gradient of f normal to the boundary, then, and according to the notation of Fig. 2.10, a fictitious point V is introduced in the flow domain and the $\partial f/\partial n$ derivative at point R is approximated by the difference

$$\left.\frac{\partial f}{\partial n}\right|_R = \frac{f_V - f_I}{1_{IV}} \quad (2.44)$$

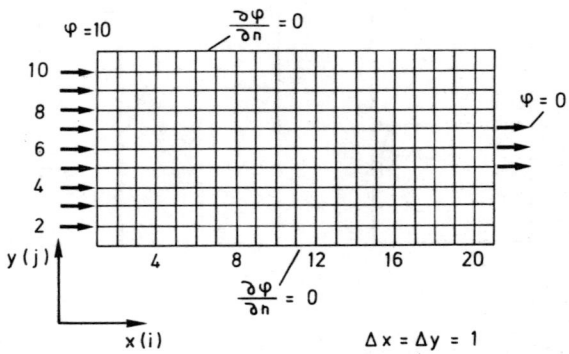

Fig. 2.11

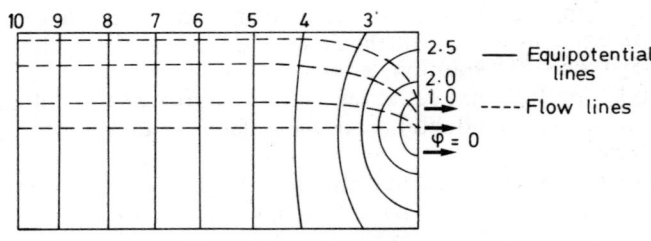

Fig. 2.12

The f_I value is computed by means of interpolation between the f_Q and f_P values. Equation 2.44 is used for the computation of the unknown f_V value.

The use of the above numerical scheme for the integration of the Laplace equation in a 2-D domain is demonstrated in the following example.

EXAMPLE 2.5

Assume the flow of an ideal fluid in the flow domain given in Fig. 2.11. The flow is from left to right.

The velocity potential $\varphi(\vec{V} = -\text{grad }\varphi)$ satisfies the Laplace equation. The φ values on the nodes of the grid discretising the flow field are to be computed and the equipotential and flow lines drawn. The boundary conditions are of mixed type and are given in Fig. 2.11. The solution of the resulting algebraic system is to be performed by Gauss–Seidel iterations.

The FORTRAN listing could have the form

```
C     SOLUTION OF LAPLACE EQN CENTRAL DIFFERENCS
      DIMENSION F(50,50)
      READ(5,5) DX, IMAX, JMAX
    5 FORMAT(F7.0,2I4)
      DO 6 I=1,IMAX
      DO 6 J=1,JMAX
    6 F(I,J)=0.
      IMAX1=IMAX-1
      JMAX1=JMAX-1
      DO 7 J=1,JMAX
    7 F(1,J)=10.
      DO 8 J=5,7
    8 F(IMAX,J)=0.
      ITER=0
  100 ITER=ITER+1
      IF (ITER.GT.1000) STOP 1
      DIFMX=0.
      DO 9 I=2,IMAX1
      DO 9 J=2,JMAX1
      TEMP=F(I,J)
      F(I,J)=(F(I,J+1)+F(I,J-1)+F(I+1,J)+
     F(I-1,J))/4.
      DIF=ABS(TEMP-F(I,J))
      IF(DIF.GT.DIFMX) DIFMX=DIF
    9 CONTINUE
      DO 10 I=2,IMAX1
      F(I,1)=F(I,2)
   10 F(I,JMAX)=F(I,JMAX1)
      DO 11 J=2,4
   11 F(IMAX,J)=F(IMAX1,J)
      DO 12 J=8,10
   12 F(IMAX,J)=F(IMAX1,J)
      IF(DIFMX.GT..0001) GO TO 100
      DO 13 J=JMAX, 1, -1
   13 WRITE(6,14)(F(I,J),I=1,IMAX)
   14 FORMAT(21F6.2)
      STOP
      END
```

Data: $DX=1$, $IMAX=21$, $JMAX=11$.

The equipotential lines deriving from the numerical solution are shown in Fig. 2.12; the lines normal to them give qualitatively the flow lines of the fluid particles.

3

Flow in closed conduits

3.1. MATHEMATICAL MODELS FOR STEADY FLOW IN PIPES AND PIPE NETWORKS

The steady flow in closed conduits either in isolation or connected to networks is quantitatively described by means of algebraic equations deriving from the physical principles of mass continuity and energy conservation along the water flow.

The mass continuity for an isolated pipe segment is expressed by the conservation of discharge along the pipe. If S_i denotes the area of cross-section, i, and V_i the mean sectional velocity, the principle of mass continuity is expressed as

$$Q = \text{const.} = S_i V_i \tag{3.1}$$

In the case of pipe networks the mass continuity is written for a node of the network in a manner analogous to the relation for current intensity in electrical circuits (Kirchoff's law)

$$\sum_{i=1}^{n} \varepsilon_i Q_i = 0 \tag{3.2}$$

where Q_i is the discharge along the branch i confluent to the node and ε_i its sign, $\varepsilon_i = \pm 1$. The algebraic sign is relatively fixed (for example $\varepsilon = +1$ for incoming discharge to the node and $\varepsilon = -1$ for outgoing discharge).

The energy conservation principle for an isolated pipe, along which a discharge Q flows, is expressed by equating the pressure head drop to the distributed and local head losses. According to the notation of Fig. 3.1, the equation is composed of terms whose dimensions are expressed in pressure head (or the energy loss per weight of discharge).

$$\Delta h_{12} = \frac{\lambda L U^2}{2gD} + \sum_i K_i \frac{U^2}{2g} - \Delta h_{\text{pump}} + \Delta h_{\text{turbine}} \tag{3.3}$$

FLOW IN CLOSED CONDUITS 49

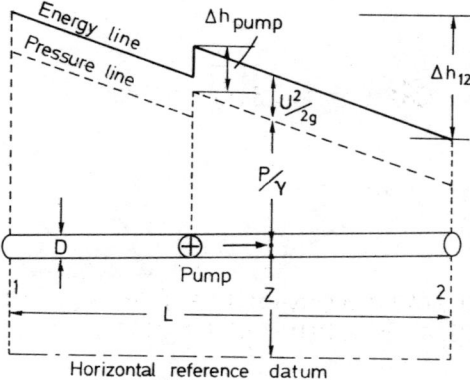

Fig. 3.1

The inclusion of a pump appears as an energy source term in the equation, while the inclusion of a turbine as an energy sink term. The local losses are expressed in terms of the local head loss coefficients K_i while the distributed energy losses introduce the Darcy loss coefficient λ, a function of the pipe wall roughness, and the Reynolds number Re. For large Re as in most water supply or irrigation projects the coefficient λ becomes independent of Re.

The Δh_{pump} term can be substituted by the expression

$$P_{\text{pump}} = \frac{\gamma Q \Delta h_{\text{pump}}}{n} \rightarrow \Delta h_{\text{pump}} = \frac{Pn}{\gamma Q} \tag{3.4}$$

where P is the pump power, n the efficiency coefficient, γ the specific weight of water ($\gamma = 10^3$ kg/m^3) and Q the discharge.

The hydraulic resistance coefficient R is defined by

$$\frac{\lambda L U^2}{2gD} = \frac{8\lambda L Q^2}{g\pi^2 D^5} = RQ^2 \quad \text{where} \quad R = \frac{8\lambda L}{g\pi^2 D^5} \tag{3.5}$$

After the drop due to the local losses (Equation 3.3), which may be incorporated into the linear losses through equivalent resistance coefficients, the head losses are expressed in terms of the discharge by,

$$\Delta h_{12} = RQ^2 + \frac{Pn}{\gamma Q} \tag{3.6}$$

It should be noted that the energy line height above a horizontal

50 FLOW IN CLOSED CONDUITS

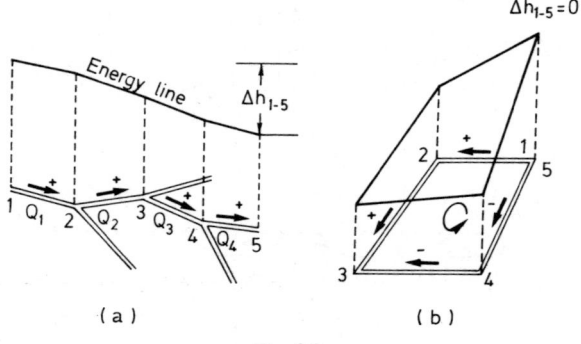

Fig. 3.2

reference datum is given by the sum (Fig. 3.1),

$$h_E = z + \frac{P}{\gamma} + \frac{U^2}{2g} \tag{3.7}$$

and that the pressure line lies below the energy line at a distance equal to $U^2/2g$

$$h_P = z + \frac{P}{\gamma} \tag{3.8}$$

For a group of connected pipes belonging to a tree-shape network, Equation 3.6 can be written for all the pipes belonging to a route (a line running along pipes without passing from the same node twice). For the pipes 1–2, 2–3, 3–4, 4–5, of the pipe system of Fig. 3.2(a), Equation 3.6 can be written as

$$\Delta h_{15} = \sum_{i=1}^{n} \varepsilon_i R_i Q_i^2 \tag{3.9}$$

The signs are fixed according to an arbitrarily chosen positive direction along the route 1–5. Such equations can be written for all the routes of a tree system.

For closed networks, the independent pipe meshes have to be determined and the energy conservation equations written for all the pipes of a mesh. As the start and end of the route in a closed network coincide (point $1 \equiv$ point 5, Fig. 3.2(b)) the energy equation takes the form

$$\sum_{i=1}^{n} \varepsilon_i R_i Q_i^2 = \Delta h_{1-5} = 0 \qquad (3.10)$$

An arbitrary positive direction has to be chosen for each mesh.

The problems posed in pipe networks are of two kinds: (1) the calculation of the discharges through the pipes when the pipe material, length and diameters are given and (2) the discharges are given and the pipe diameters are to be computed. From the computed discharges or diameters, energy and pressure lines are estimated.

Various limitations arise during calculations, such as upper and lower velocity limits, minimum or maximum pressure heads at the nodes, and pipe diameters commercially available. Also, optimisation problems with respect to construction costs or the available pressure head have to be solved. When electronic computers are available simple trial-and-error procedures can be applied to determine the optimum economic and hydraulic conditions. Iterative procedures have been developed, such as the Labee method for open pipe systems applicable to irrigation projects.

3.2. STEADY FLOW IN PIPE NETWORKS (THE HARDY-CROSS METHOD)

The computation of either the discharges or diameters of tree-shaped pipe systems is straightforward as the equations for mass continuity and energy conservation are uncoupled and the solution can proceed usually from downstream to upstream pipes. A problem arises with the computation of discharges along the branches of pipe networks when the diameters are given.

The available relations can be expressed as follows:

$$\text{Mass continuity for each node,} \quad \sum_{i=1}^{n} \varepsilon_i Q_i = 0 \qquad (3.11)$$

where n is the number of pipes contributing to the node.

$$\text{Energy conservation for each mesh,} \quad \sum_{i=1}^{k} \varepsilon_i R_i Q_i^2 = 0 \qquad (3.12)$$

where k is the number of pipes around the mesh.

Equations 3.11 and 3.12 are coupled to a system of nonlinear algebriac equations. A classical method of solution by successive approximations is known as the Hardy-Cross method. Arbitrary discharge values for each pipe are initially chosen in such a way as to satisfy the mass continuity relations at all the nodes. This selection can be easily done and the selected values form the first approximation to the solution vector $Q_i^{(1)}$. The energy relations are subsequently applied to each mesh. Due to the arbitrariness of the first selection these equations are not satisfied, and residuals appear for each mesh. The correction of discharges on each mesh is based on these residuals. Suppose that the head residual on the mesh A is

$$R_A = \sum_{i=1}^{n} \varepsilon_i R_i Q_i^2 \qquad (3.13)$$

If the discharges Q_i are corrected by $\varepsilon_i \Delta Q$ they will become $Q_i + \varepsilon_i \Delta Q$ and the residual will vanish, i.e.

$$\sum \varepsilon_i (Q_i + \varepsilon_i \Delta Q)^2 R_i = 0 \approx \sum_{i=1}^{n} R_i \varepsilon_i Q_i^2 + \sum_{i=1}^{n} 2 Q_i \Delta Q R_i \qquad (3.14)$$

This approximate equation can be solved for ΔQ,

$$\Delta Q = \frac{-\sum R_i \varepsilon_i Q_i^2}{2 \sum R_i Q_i} \qquad (3.15)$$

The $\varepsilon_i \Delta Q$ correction does not modify the mass continuity at all the nodes (insured from the beginning of the iterative solution). After a number of successive corrections for all the network meshes, convergence to the final discharge values satisfying both continuity and conservation principles is usually achieved.

The computer programming of the method is simple. The hydraulic resistance coefficients are initially introduced or computed from length, diameter and Darcy coefficient values. The first discharge values satisfying mass continuity are also introduced as initial conditions. A matrix A with elements a_{ik} is subsequently introduced. The number of columns is equal to the number of pipes of the network and the number of rows equal to the number of independent meshes. The a_{ik} element takes the following values: $a_{ik} = +1$ or -1 if the mesh i contains pipe k with positive or negative flow direction, respectively (according to the selected positive mesh direction) and $a_{ik} = 0$ if the mesh i does not contain pipe k.

The discharge corrections are computed for each mesh by means of Equation 3.15. The most recent discharge values are used each time. A change of flow direction in a pipe k during the iterations affects the matrix as the a_{ik} elements change sign for all i and specific k (pipe number).

FLOW IN CLOSED CONDUITS 53

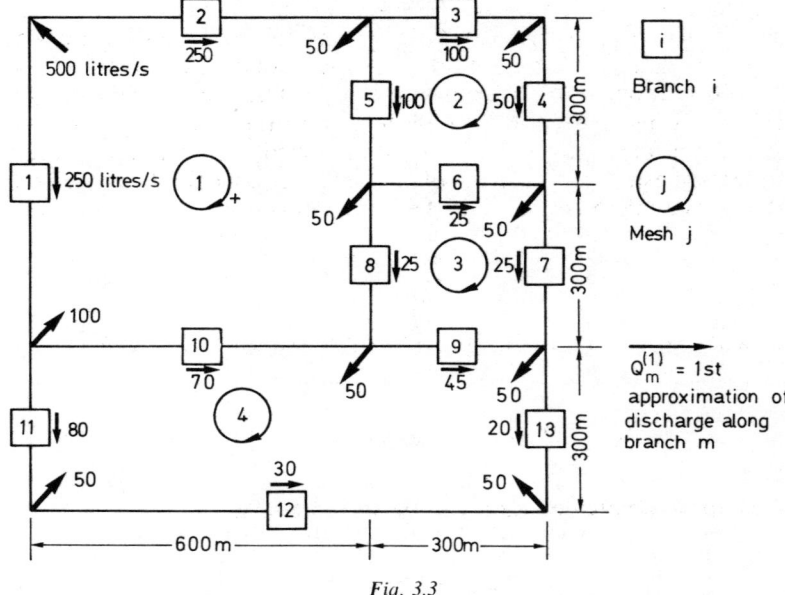

Fig. 3.3

Convergence is reached when the maximum ΔQ value of all the meshes becomes less than a preimposed test value. The losses along the pipes and the pressure heads at all the nodes can be subsequently computed on the basis of the known discharge.

The computed velocity values and available heads must lie within preimposed limits. Otherwise proper diameter modifications are done and the problem is resolved. The following example demonstrates the organisation of the solution for a 4-mesh network.

EXAMPLE 3.1

To compute the discharge in the branches of the network of Fig. 3.3. The diameters, lengths and Darcy coefficients of the pipes are given (Table 3.1), together with the discharge extracted at the nodes for water supply purposes. If the pressure head E at the network entrance is 40 m what are the pressure heads at the rest of the nodes?

We proceed as follows:

(1) The resistance coefficients R_i are computed for each branch of the network.

(2) The initial values of discharges are fixed arbitrarily such as to

Table 3.1

i	l_i (m)	D_i (m)	λ_i	R_i
1	600	0.4	0.02	96.83
2	600	0.4	0.02	96.83
3	300	0.3	0.03	306.00
4	300	0.3	0.03	306.00
5	300	0.3	0.03	306.00
6	300	0.2	0.04	3098.00
7	300	0.2	0.04	3098.00
8	300	0.2	0.04	3098.00
9	300	0.3	0.03	306.00
10	600	0.3	0.03	612.00
11	300	0.3	0.03	306.00
12	900	0.2	0.04	9295.5
13	300	0.2	0.04	3098.00

Table 3.2

	1	2	3	4	5	6	7	8	9	10	11	12	$13 = i_{max}$
1	−1	+1	0	0	+1	0	0	+1	0	−1	0	0	0
2	0	0	+1	+1	−1	−1	0	0	0	0	0	0	0
3	0	0	0	0	0	+1	+1	−1	−1	0	0	0	0
4	0	0	0	0	0	0	0	0	+1	+1	−1	−1	−1

satisfy the continuity conditions at the nodes. The signs are given according to the positive direction around each mesh (Table 3.2).
(3) The branch–node matrix is formed.
(4) The solution by the Hardy-Cross method is programmed.

```
C     H.CROSS SOLUTION OF PIPE NETWORK
      DIMENSION R(40),Q(40),M(40,40),U(40),DH(40)
     1D(40)
      READ(5,1) TEST,IMAX,KMAX
    1 FORMAT(F7.0,2I4)
      READ(5,10)(Q(I),I=1,IMAX)
      READ(5,10)(D(I),I=1,IMAX)
      DO 11 K=1,KMAX
   11 READ(5,12)(M(K,I),I=1,IMAX)
      READ(5,10)(R(I),I=1,IMAX)
   12 FORMAT(13I4)
   10 FORMAT(10F7.0)
      ITER=0
   40 ITER=ITER+1
      IF (ITER.GT.500) STOP 1
      DIFMX=0.
      DO 20 K=1, KMAX
      SUM1=0.
```

```
        SUM2=0.
        DO 25 I=1, IMAX
        T=M(K,I)
        SUM1=SUM1+T*R(I)*Q(I)**2
 25     SUM2=SUM2+2.*ABS(T)*Q(I)*R(I)
        DQ=-SUM1/SUM2
        IF(ABS(DQ).GT.DIFMX) DIFMX=ABS(DQ)
        DO 30 I=1, IMAX
        Q(I)=Q(I)+M(K,I)*DQ
        IF(Q(I).GT.O.) GO TO 30
        Q(I)=-Q(I)
        DO 35 KK=1, KMAX
 30     CONTINUE
 20     CONTINUE
        IF(DIFMX.GT.TEST) GO TO 40
        DO 70 I=1, IMAX
        U(I)=Q(I)*1.273/D(I)**2
 70     DH(I)=R(I)*Q(I)**2
        WRITE(6,75) ITER
 75     FORMAT(//I8//)
        WRITE(6,45)(Q(I), I=1, IMAX)
        WRITE(6,45)(U(I), I=1, IMAX)
        WRITE(6,45)(DH(I), I=1, IMAX)
        DO 65 K=1,KMAX
 65     WRITE(6,55)(M(K,I),I=1, IMAX)
 45     FORMAT(10F10.4)
 55     FORMAT(10I6)
        STOP
        END
```

The final discharge values along the branches (for a convergence criterion TEST = 0.1) and the calculated pressure heads at the nodes are depicted in Fig. 3.4.

3.3. NON-STEADY FLOW. WATER HAMMER

Unsteady flow in closed conduits becomes very important in the case of sudden changes of discharge due to the interruption of a pump operation, or the closing or opening of a vane. These variations create pressure waves propagating with alternating sign (high pressure or low pressure) along a pipe. The computational goal is usually to find the extreme pressure values in order to check the safety of the conduit.

Assuming elastic conduit walls and compressible fluid the velocity of an elastic wave along the pipe is found to be

$$c = \sqrt{\left\{ \frac{1}{\rho\left(\frac{1}{K} + \frac{D}{Ee}\right)} \right\}} \tag{3.16}$$

56 FLOW IN CLOSED CONDUITS

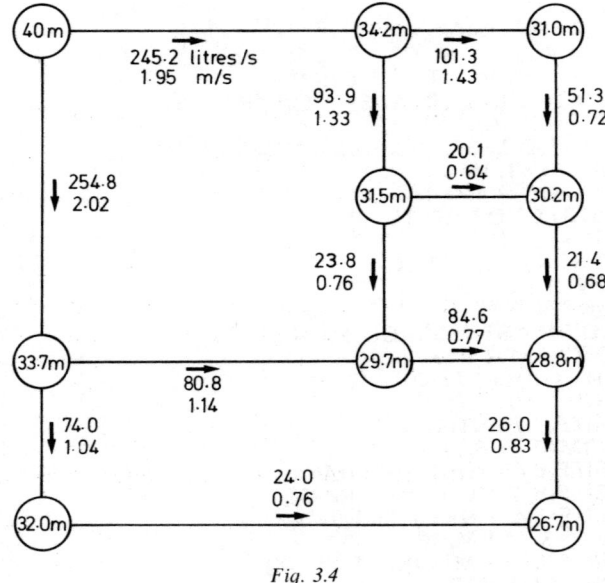

Fig. 3.4

where ρ is the water density, K the water modulus of elasticity ($K = 2 \times 10^8$ $k\rho/\text{m}^2$), D the pipe diameter, e the walls thickness, and E the modulus of elasticity of the pipe material ($E = 2 \times 10^{10}$ $K\rho/\text{m}^2$ for steel pipes).

Approximating the pipe body by a series of rings (negligible Poisson ratio) with uniform internal pressure, we can correlate the relative pressure inside the pipe to the pipe diameter,

$$H = \frac{2c^2}{g} \frac{r - r_0}{r} \qquad (3.17)$$

where r is the radius under pressure head H and r_0 the radius under pressure head equal to zero (absolute pressure = 1 atm).

The mathematical model is formed using the quantitative expression for the principle of mass continuity and force equilibrium, written with respecto the unknown functions H, V (pressure head and water velocity) as:

$$\frac{\partial V}{\partial t} + V \frac{\partial V}{\partial x} + g \frac{\partial H}{\partial x} = -\frac{2\tau_0}{r} \qquad (3.18)$$

FLOW IN CLOSED CONDUITS

$$\frac{\partial H}{\partial t} + V\frac{\partial H}{\partial x} + \frac{c^2}{g}\frac{\partial V}{\partial x} = 0 \tag{3.19}$$

where τ_0 is the wall shear stress, which can be expressed as

$$\tau_0 = KV|V| \tag{3.20}$$

where

$$K = \lambda\rho/8 \tag{3.21}$$

a friction coefficient depending on the pipe diameter and roughness.

During the numerical integration of Equations 3.18 and 3.19 using finite difference methods, some difficulties arise on the upstream and downstream boundaries where the H, V values, on and outside these boundaries are needed. This difficulty can be overcome using the properties of the corresponding characteristic curves. A considerable simplification is achieved by neglecting the non-linear terms $V\partial V/\partial x$ and $V\partial H/\partial x$.

As the frictional losses are usually small, due to the short duration of the phenomenon and the moderate velocity values, the system of Equations 3.18 and 3.19 can be simplified to the final form,

$$\frac{\partial V}{\partial t} + g\frac{\partial H}{\partial x} = 0 \tag{3.22}$$

$$\frac{\partial H}{\partial t} + \frac{c^2}{g}\frac{\partial V}{\partial x} = 0 \tag{3.23}$$

The model is completed by the initial and boundary conditions.

The velocity and pressure heads are given before the initiation of the velocity variations. The boundary conditions are usually either of the reservoir type, where the pressure head is constant (equal to the hydrostatic pressure) or of the vane type. In the case of a suddenly closed vane the velocity is zero and in the case of a slowly closing or opening vane the continuity principle leads to the condition,

$$V = \frac{S}{S_0}\sqrt{(2gH)} \tag{3.24}$$

where $S = S(t)$ is the flow section, S_0 the initial vane opening and H the pressure head.

The numerical integration of the system describing the propagation of pressure and discharge waves along the pipe, with or

58 FLOW IN CLOSED CONDUITS

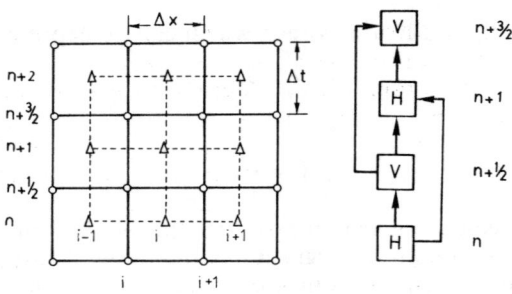

Fig. 3.5

without simultaneous energy dissipation due to friction, can be done by the developed FD schemes.

A simple explicit scheme will be used here directly applied to a staggered grid. The grid is characterised by temporal and spatial eccentricity. The pipe is discretised in characteristic sections. The velocity values are computed for each section while the pressure heads refer to the reaches between two sections. The H and V computations are performed at different time levels. The H^n, H^{n+1} ... and $V^{n+\frac{1}{2}}$, $V^{n+\frac{3}{2}}$ ... values are interchangeably computed. The grid in x–t space and the integration procedure are schematically given in Fig. 3.5. The approximation of Equations 3.22 and 3.23 by difference equations leads to,

$$\frac{V_i^{n+3/2} - V_i^{n+1/2}}{\Delta t} + g \frac{H_i^{n+1} - H_{i-1}^{n+1}}{\Delta x} = 0 \qquad (3.25)$$

$$\frac{H_i^{n+1} - H_i^n}{\Delta t} + \frac{c^2}{g} \frac{V_{i+1}^{n+1/2} - V_i^{n+1/2}}{\Delta x} = 0 \qquad (3.26)$$

where i and n are space and time indices respedtively.

An application for an isolated pipe starting from a reservoir and terminated by a closing vane is given in the following example. The closing time is smaller than $2L/c$ where L is the pipe length.

EXAMPLE 3.2

To compute the pressure head variations on the downstream end of a pipe $L = 6000$ m long with $D = 0.5$ m, $e = 4$ mm, $H_0 = 5$ m and $V_0 =$

FLOW IN CLOSED CONDUITS 59

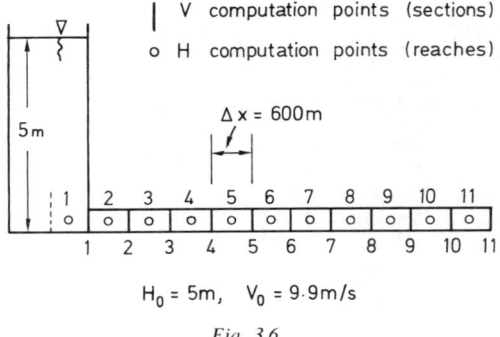

$H_0 = 5m$, $V_0 = 9.9 m/s$

Fig. 3.6

9.9 m/s. The discharge is interrupted through a downstream vane at $t = 2$ s with a linear decrease of the flow section. Frictional losses are negligible.

A general plan of the reservoir-pipe system and the pipe discretisation is included in Fig. 3.6. The solution is performed by means of Equations 3.25 and 3.26. At the upstream end the pressure head, equal to the reservoir depth, is constant. At the downstream end the boundary condition relates velocity and flow section. The wave speed is computed from Equation 3.16 and is found to be $c = 2980$ m/s. The integration time step Δt is taken equal to 0.1 s so that the Courant condition is satisfied.

$$\frac{c\Delta t}{\Delta x} = 0.496 < 1 \qquad (3.27)$$

The solution is programmed in FORTRAN as follows:

```
C     WATER HAMMER FINITE DIFF SOLUTION
      DIMENSION V(100), H(100)
      READ(5,1)DX,DT,C,HO,VO,TOL,IMAX
    1 FORMAT(6F7.0,I4)
      IMAX1 = IMAX−1
      DO 2 I=1, IMAX
      V(I) = VO
    2 H(I) = HO
      T=0.
      N=0
  100 T=T+DT
      N=N+1
      IF(T.GT.TOL) GO TO 3
      SR=(TOL−T)/TOL
      GO TO 4
```

```
    3   SR=0.
    4   CONTINUE
        V(IMAX)=SR*SQRT(2.*9.81*HO)
        DO 5 I=2,IMAX1
    5   V(I)=V(I)-9.81*DT*(H(I)-H(I-1))/DX
        DO 6 I=2,IMAX1
    6   H(I)=H(I)-DT*C**2/9.81*(V(I+1)-V(I))/DX
        WRITE(6,11) T
   11   FORMAT (//F10.3//)
        WRITE(6,12)(V(I),I=1,IMAX)
        WRITE(6,12)(H(I),I=1,IMAX)
   12   FORMAT(12F10.4)
        IF(N.LT.200) GO TO 100
        STOP
        END
```

For data values $DX = 600$ m, $DT = 0.1$ s, $HO = 5$ m, $VO = 9.9$ m/s, $TOL = 2$ s (closure time) and $IMAX = 12$ sections. The pressure variation at the downstream end is graphically presented in Fig. 3.7.

The accuracy achieved by the numerical solution is satisfactory as the analytical solution gives for the maximum excess pressure

$$\Delta H = \frac{c \Delta V}{g} = 3010 \text{ m} \qquad (3.28)$$

a value differing very little from the numerically computed one. The method does not contain any numerical diffusion as the maximum and minimum pressure values are periodically repeated in the absence of frictional losses.

The linearised model, Equations 3.22 and 3.23, applied here presents no difficulty on the boundaries as the non-linear $V \partial V/\partial x$ and $V \partial V/\partial x$ terms are dropped. The inclusion of these terms for rapidly varying velocity fields, as in the case of very deformable pipe walls and in the case of open conduits, would introduce some complications.

The property of characteristics is used here and in Chapter 4 at an introductory level as this facilitates both the understanding of the wave propagation mechanism and the computational procedures.

If Equation 3.23 is multiplied by g/c and successively added and subtracted from Equation 3.22 the following relations result:

$$\frac{\partial}{\partial t}\left(V + \frac{gH}{c}\right) + c \frac{\partial}{\partial x}\left(V + \frac{gH}{c}\right) = 0 \qquad (3.29)$$

$$\frac{\partial}{\partial t}\left(V - \frac{gH}{c}\right) - c \frac{\partial}{\partial x}\left(V - \frac{gH}{c}\right) = 0 \qquad (3.30)$$

FLOW IN CLOSED CONDUITS 61

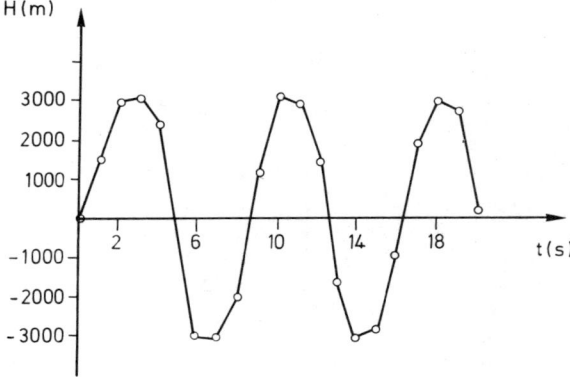

Fig. 3.7

If the dx/dt derivatives are replaced by $+c$ and $-c$ respectively in Equations 3.29 and 3.30 they take the form

$$\frac{d}{dt}\left(V+\frac{gH}{c}\right)=0, \quad \text{along} \quad \frac{dx}{dt}=c \qquad (3.31)$$

$$\frac{d}{dt}\left(V-\frac{gH}{c}\right)=0, \quad \text{along} \quad \frac{dx}{dt}=-c \qquad (3.32)$$

where d/dt denotes the total or material derivative.

The physical meaning of Equations 3.31 and 3.32 is that the characteristic lines are straight lines of slope $+c$ and $-c$ and the $V\pm(gH/c)$ values are kept constant along those lines. Their numerical integration can be realised on the characteristic lines or, for reasons of easier geometric description of the flow domain, on an orthogonal grid established in the x–t plane. A simple explicit procedure of integration on the orthogonal grid will be presented below.

The notation to be used is presented in Fig. 3.8. The characteristics AL, AR with slopes $+c$ and $-c$, respectively, pass through the point A where the values of V and H are to be computed. If the inequality $\Delta x > c\Delta t$ is satisfied, points L and R lie between points $i-1, i$ and $i, i+1$, respectively. The sums $V+(gH/c)$, $V-(gH/c)$ are known to be constant along these lines.

The values of V, H on L and R (V_L, H_L, V_R, H_R) can be easily computed by means of linear interpolation from the known values $V_{i-1}^n, V_i^n, V_{i+1}^n, H_{i-1}^n, H_i^n, H_{i+1}^n$. On this basis, the following algorithm can be formulated:

62 FLOW IN CLOSED CONDUITS

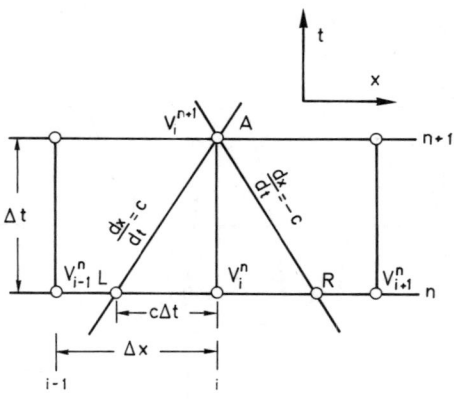

Fig. 3.8

(1) Computation of V_L, H_L, V_R, H_R:

$$V_L = V_i^n + (V_{i-1}^n - V_i^n)c\Delta t/\Delta x \qquad (3.33)$$

$$V_R = V_i^n + (V_{i+1}^n - V_i^n)c\Delta t/\Delta x \qquad (3.34)$$

$$H_L = H_i^n + (H_{i-1}^n - H_i^n)c\Delta t/\Delta x \qquad (3.35)$$

$$H_R = H_i^n + (H_{i+1}^n - H_i^n)c\Delta t/\Delta x \qquad (3.36)$$

(2) Computation of the auxiliary quantities AL, AR by means of Equations 3.33–3.36:

$$AL = V_L + \frac{g}{c} H_L \qquad (3.37)$$

$$AR = V_R - \frac{g}{c} H_R \qquad (3.38)$$

(3) Computation of V_i^{n+1}, H_i^{n+1} from Equations 3.31 and 3.32:

$$V_i^{n+1} = (AL + AR)/2 \qquad (3.39)$$

$$H_i^{n+1} = (AL - AR)/(2*c/g) \qquad (3.40)$$

The **H** values are known on the upstream boundary and the **V** values can be computed from the AR characteristic by means of the relation

$$V_1^{n+1} = \frac{g}{c} H_1^{n+1} + AR_1 \tag{3.41}$$

The velocity values are known on the downstream boundary and the H values can be computed from the AL characteristic

$$H_{i\max}^{n+1} = (AL_{i\max} - V_{i\max}^{n+1})c/9.81 \tag{3.42}$$

The algorithm can be programmed in FORTRAN as follows:

```
C     WATER HAMMER SOLUTION BY CHARACTERISTICS
      DIMENSION AL(40), AR(40), V(40), H(40)
      READ(5,1) DX,DT,C,HO,VO,TOL,IMAX
  1   FORMAT(6F7.0,I4)
      IMAX1=IMAX-1
      DO 2 I=1,IMAX
      V(I)=VO
  2   H(I)=HO
      N=0
      T=0.
 100  T=T+DT
      N=N+1
      DO 3 I=2,IMAX
  3   AL(I)=V(I)+9.81/C*H(I)+(-V(I)-9.81/C*H(I)+
     1   V(I-1)+9.81/C*H(I-1))*C*DT/DX
      DO 5 I=1,IMAX1
  5   AR(I)=V(I)-9.81/C*H(I)+(-V(I)+9.81/C*H(I)+
     1   V(I+1)-9.81/C*H(I+1))*C*DT/DX
      IF (T.GT.TOL) GO TO 33
      SR=(TOL-T)/TOL
      GO TO 34
 33   SR=0.
 34   CONTINUE
      H(IMAX)=(AL(IMAX)-V(IMAX))*C/9.81
      H(1)=HO
      V(1)=9.81/C*H(1)+AR(1)
      V(IMAX)=SR*SQRT(2.*9.81*HO)
      DO 6 I=2,IMAX1
      V(I)=(AL(I)+AR(I))/2.
  6   H(I)=(AL(I)-AR(I))*C/2./9.81
      WRITE(6,10) T
 10   FORMAT(//5X, 'TIME',F10.3//)
      WRITE(6,12)(H(I),I=1,IMAX)
      WRITE(6,12)(V(I),I=1,IMAX)
 12   FORMAT(11F10.4)
      IF (N.LT.200) GO TO 100
      STOP
      END
```

64 FLOW IN CLOSED CONDUITS

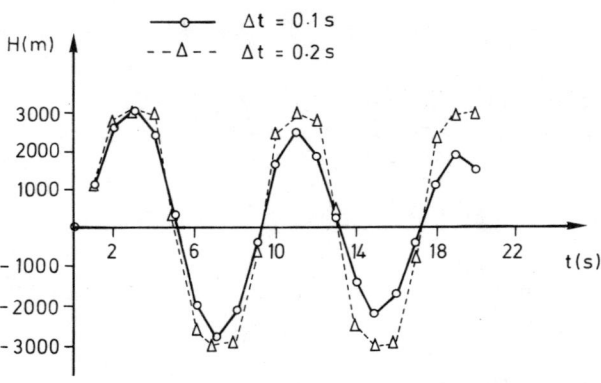

Fig. 3.9

EXAMPLE 3.3

Repeat Example 3.2 using for the numerical integration the properties of characteristics. The pressure variation at the downstream end is to be computed for $t = 0.1$ s and $t = 0.2$ s.

The application of the above program to the previous data values results in the velocity and pressure head values at the characteristic sections of the pipe. The pressure variation at the downstream end is graphically represented in Fig. 3.9.

It is evident that the explicit scheme using the properties of characteristics contains some inherent numerical dissipation not apparent in the previous solution of Example 3.2. The dissipation seems to vary in inverse proportion to the Δt value. As the value of $c\Delta t/\Delta x$ tends to unity the numerical dissipation decreases. The maximum pressure tends to that found analytically, i.e. 3010 m.

4
Open channel flow

4.1. MATHEMATICAL MODELS FOR NON-STEADY FLOW IN OPEN CHANNELS

Unsteady flow with a free surface forms one of the most interesting areas of hydraulics. Long waves and tidal flow in estuaries, the propagation of floods along natural water-courses, transient flow in irrigation canals due to discharge and level fluctuations are only some examples that fall into this category.

The mathematical model of flow in an open channel with variable cross-section, extending in one dimension in the x direction, contains as unknown functions the mean velocity over a cross section $V = V(x, t)$ and the flow depth $h = h(x, t)$, measured from the lowest part of the cross section to the free surface. It can be synthesized from the quantification of the basic principles of mass continuity and conservation of momentum between two cross-sections. According to the notation of Fig. 4.1 it can take the form

$$B \frac{\partial h}{\partial t} + \frac{\partial}{\partial x}(AV) = 0 \qquad (4.1)$$

$$\frac{\partial V}{\partial t} + V \frac{\partial V}{\partial x} - g(S_0 - S_f) + g \frac{\partial h}{\partial x} = 0 \qquad (4.2)$$

The slope of the energy line S_f can be approximated, even in the case of unsteady flow, by means of semi-empirical formulae valid for steady flow (the Manning or Chézy equations),

$$S_f = \frac{V^2}{K^2 R^{4/3}} = \frac{V^2}{C^2 R} \qquad (4.3)$$

where K and C are the Manning and Chézy friction coefficients, respectively, and R the hydraulic radius, $R = A/P$. In the case of a channel of large width, B, in comparison to depth, Equation 4.1 can take the form,

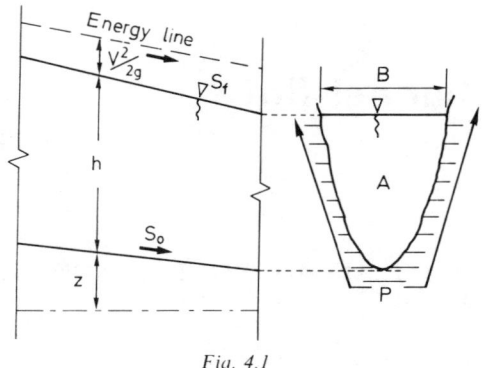

Fig. 4.1

$$\frac{\partial h}{\partial t} + \frac{1}{B}\frac{\partial}{\partial x}(BVh) = 0 \qquad (4.4)$$

and R in Equation 4.3 can be approximated by h. When the channel width is constant, Equation 4.4 becomes

$$\frac{\partial h}{\partial t} + \frac{\partial}{\partial x}(Vh) = 0 \qquad (4.5)$$

Sometimes it is preferable to write the equations in terms of the discharge function $Q(x, t)$ and the depth h. From Equations 4.1 and 4.2, with appropriate manipulation, we can derive,

$$B\frac{\partial h}{\partial t} + \frac{\partial Q}{\partial x} = 0 \qquad (4.6)$$

$$\frac{\partial Q}{\partial t} + \frac{\partial}{\partial x}\left(\frac{Q^2}{A}\right) + gA\frac{\partial h}{\partial x} - gA(S_0 - S_f) = 0 \qquad (4.7)$$

The system of Equations 4.1 and 4.2, or its equivalent, is completed by the appropriate boundary conditions. Their form depends on the nature of the upstream and downstream boundaries, either natural or artificially defined.

(1) Upstream boundary of given discharge hydrograph. On this boundary, the discharge function (flood hydrograph) is supplied

$$Q = A(h)V \qquad (4.8)$$

If the equations are in the form of Equations 4.6 and 4.7, the hydrograph is used directly, otherwise only the product of the two unknown functions h, V is known.

(2) Reflection boundary. On this boundary both velocity and discharge are suppressed.

(3) Fluctuating surface boundary. On this boundary the fluctuation of the water depth $h(t)$ is prescribed. Such boundaries are, for example, the downstream ends of estuaries flowing into tidal seas, and some boundaries in irrigation canals.

(4) Free transmission (no reflection) boundary, i.e. assumed no back reflection of velocity or depth perturbation. If no more detail is known, the following correlation between velocity and depth can be used,

$$V = \sqrt{gh} \to \delta Q = \delta A \sqrt{gh} \qquad (4.9)$$

(5) Boundary with known, prescribed relation between discharge and depth, $Q = Q(h)$, as in most cases a river flow where depth–discharge measurements can be correlated.

The simplification of the non-linear equations can be done on grounds of the relative magnitude and importance of the several terms, for each special flow situation. The most common simplifications are the dropping of the non-linear spatial acceleration (advective) term, $V(\partial V/\partial x)$, and the term for frictional losses, gS_f.

In the case of flood routing along a river there are simplified models where instead of Equation 4.7, the steady flow equation relating Q and h can be used. These and other simplifications can be done only after a detailed investigation of their validity, and they should be checked after the computations by comparing the analytical solution or in situ measurements.

4.2. NUMERICAL SOLUTIONS FOR LONG WAVE PROPAGATION

From a close examination of Equations 4.1 and 4.2, it can be seen that the mathematical model is composed of partial differential equations of mixed form. They contain a hyperbolic part, consisting of the temporal and spatial acceleration terms and the pressure gradient term (describing a signal propagation), and a parabolic part consisting of the temporal acceleration and the frictional term (describing dissipation due to frictional processes).

The numerical methods described in Chapter 2 for the solution of common partial differential equations in hydraulics can be transferred and applied here without modification. Otherwise, from

68 OPEN CHANNEL FLOW

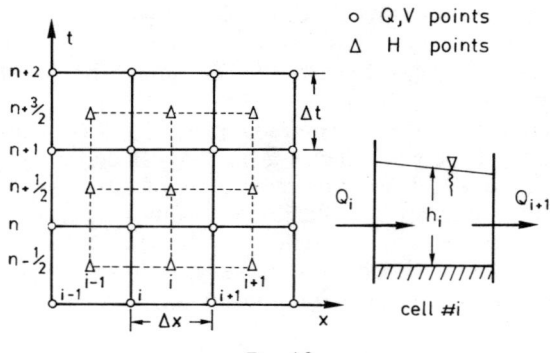

Fig. 4.2

the similarity between the form of the mathematical model of unsteady flow in open channels, to that in closed conduits, it can be concluded that their solution algorithms are similar.

An explicit and an implicit finite difference scheme will be presented in this section together with an integration procedure based on the existence of characteristic curves for Equations 4.1 and 4.2.

The integration of the model, in the form of Equations 4.6 and 4.7, by the method of finite differences, is done on the same staggered (non-centred) grid described in Section 3.3, as this is convenient for the particular boundary conditions of the given discharge hydrograph described in Fig. 4.2. This consists of cells, with the continuity and momentum conservation equations expressed in terms of the velocity values at the nodes and the water depth values at the center of each cell.

According to the notation of Fig. 4.2, Equation 4.2 is approximated by means of finite differences in the following way:

$$\frac{h_i^{n+1/2} - h_i^{n-1/2}}{\Delta t} + \frac{1}{B_i}\left(\frac{Q_{i+1}^n - Q_i^n}{\Delta x}\right) = 0 \quad (4.10)$$

Equation 4.10 can be solved directly for $h_i^{n+\frac{1}{2}}$ (explicit scheme). The momentum equation, in the case of a rectangular uniform cross-section (A becomes h in equation 4.7 as B is constant), can be approximated through an explicit or an implicit finite difference scheme.

(1) Explicit scheme

After the introduction of the intermediate variables,

$$hh = (h_i^{n+1/2} + h_i^{n-1/2} + h_{i-1}^{n+1/2} + h_{i-1}^{n-1/2})/4 \tag{4.11}$$

$$S_f = (Q_i^n)^2/(K^2 hh^{10/3}) \tag{4.12}$$

$$hh_2 = (h_i^{n+1/2} + h_i^{n-1/2})/2 \tag{4.13}$$

$$hh_1 = (h_{i-1}^{n+1/2} + h_{i-1}^{n-1/2})/2 \tag{4.14}$$

$$QQ_2 = (Q_{i+1}^n + Q_i^n)/2 \tag{4.15}$$

$$QQ_1 = (Q_i^n + Q_{i-1}^n)/2 \tag{4.16}$$

Equation 4.7 can be written (if the time derivative is replaced by a difference of the 'Lax' type) as follows:

$$Q_i^{n+1} = \frac{(QQ_2 + QQ_1)}{2} - \Delta t \frac{\left(\dfrac{QQ_2^2}{hh_2} - \dfrac{QQ_1^2}{hh_1}\right)}{\Delta x}$$

$$+ 9.81 \Delta t \frac{hh(hh_2 - hh_1)}{\Delta x} + 9.81 * hh(S_f - S_0)\Delta t \tag{4.17}$$

(2) Implicit scheme

Defining the temporary variables,

$$QQ_1 = (Q_i^{n+1} + Q_{i-1}^{n+1} + Q_i^n + Q_{i-1}^n)^2/(16 h_{i-1}^{n+1/2}) \tag{4.18}$$

$$QQ_2 = (Q_{i+1}^{n+1} + Q_{i+1}^n + Q_i^{n+1} + Q_i^n)^2/(16 h_i^{n+1/2}) \tag{4.19}$$

$$S_f = (Q_i^{n+1} + Q_i^n)/2K^2/\{(h_i^{n+1/2} + h_{i-1}^{n+1/2})/2\}^{10/3} \tag{4.20}$$

Equation 4.7 can be written as

$$Q_i^{n+1} = Q_i^n - \Delta t(QQ_2 - QQ_1)/\Delta x - 9.81 \Delta t/2\Delta x \times$$
$$\times \{(h_i^{n+1/2})^2 - (h_{i-1}^{n+1/2})^2\} -$$
$$- 9.81 \Delta t(h_i^{n+1/2} + h_{i-1}^{n+1/2})/2(S_f - S_0) \tag{4.21}$$

It is to be pointed out, that the finite differences schemes of Equation 4.17 and 4.21 are centred with respect to time and space. This implies

70 OPEN CHANNEL FLOW

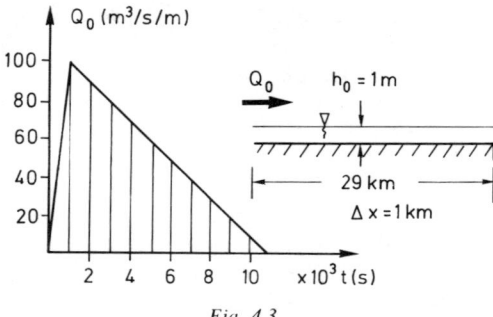

Fig. 4.3

that Equation 4.21 introduces some trailing effects (numerical dispersion) while in the case of Equation 4.17 the 'Lax' type difference introduces numerical diffusion proportional to $\Delta x^2/2\Delta t$ such that the trailing effect becomes undetectable.

Another characteristic of both the schemes is the approximation of the non-linear term $A(\partial A/\partial x)$ by means of the expression

$$A\frac{\partial A}{\partial x} \approx \frac{\left(\frac{A_{i+1}+A_i}{2}\right)^2 - \left(\frac{A_i+A_{i-1}}{2}\right)^2}{2\Delta x} \qquad (4.22)$$

which was successfully applied for the solution of Burger's equation and the description of propagation of fronts.

In the following example both the explicit and the implicit schemes are used for the description of the propagation of a flood wave along a channel of uniform rectangular cross-section.

EXAMPLE 4.1

The initial depth in an horizontal channel ($S_0 = 0$) of rectangular section is $h_0 = 1$ m. The channel has a length of 29 km and ends in a step (downstream non-reflection boundary). The initial discharge $Q_0 = 0$. The given upstream flood hydrograph is described in Fig. 4.3. The Manning friction coefficient is estimated to be $K = (1/n) = 40$ m$^{1/3}$/s, where K and n are Manning friction coefficients.

The flood arrival times, the times of maximum water depth, and the maximum water depth along the channel are to be computed by means of the described finite difference schemes. The water surface profile is to be drawn 1000 s and 2000 s after the beginning of the flood.

The flow field is first discretised in reaches of length $\Delta x = 1000$ m (number of cross sections, $i_{max} = 30$). The time step is taken equal to $\Delta t = 10$ s. The integration through an explicit scheme must satisfy the known Courant criterion $c\Delta t/\Delta x < 1$, where c is the speed of propagation of the long waves in the channel expressed by

$$c = \sqrt{(gh)} \qquad (4.23)$$

As the maximum depth, h_{max}, leading to c_{max} and Δt_{min}, is not known before hand, a conservative value $\Delta t = 10$ s is selected initially and can be corrected subsequently according to the h values. As will be seen, the time step value influences the results of the explicit 'Lax' type scheme containing numerical diffusion inversely proportional to Δt.

The introduction of the upstream discharge hydrograph is done through a series of Q values separated by a time interval $\Delta t' = 1000$ s (Fig. 4.3). At the first cross-section the $Q(1)$ values is given. At the last cross-section the $Q(IMAX)$ value is related to the mean depth along the last cell through Equation 4.9.

$$Q(IMAX) = \{H(IMAX-1) - HO\}\sqrt{\{gH(IMAX-1)\}} \qquad (4.24)$$

where HO is the initial depth.

The computational procedures described by Equations 4.17, 4.10 and 4.24 are programmed in FORTRAN as follows:
follows:

```
C     EXPLICIT FD SOLUTION OF FLOOD PROPAGATION
C     IN OPEN CHANNEL
      DIMENSION Q2(100), Q1(100),H2(100),H1(100)
     1,QQ(100)
      READ(5,5) IMAX,DT,DX,CMAN,HO,DDT
    5 FORMAT(I4,5F7.0)
      IMAX1=IMAX-1
      READ(5,6)(QQ(I), I=1,21)
    6 FORMAT(10F7.0)
      DO 7 I=1,IMAX
      H2(I)=HO
      H1(I)=HO
      Q2(I)=0.
    7 Q1(I)=0.
      T=0.
      N=0
  100 T=T+DT
      N=N+1
      DO 10 I=1,IMAX1
   10 H2(I)=H(1)-Q1(I+1)-Q1(I))/DX*DT
      L=INT(T/DDT+1)
      Q2(1)=QQ(L)+(QQ(L+1)-QQ(L))*(T-(L-1)*
```

72 OPEN CHANNEL FLOW

```
      1   DDT)/DDT
          DO 8 I=2,IMAX1
          HH=(H2(I)+H1(I)+H1(I-1)+H2(I-1))/4.
          SF=Q1(I)**2/CMAN**2/HH**3.333
          HH2=(H2)(I)+HI(I))/2.
          HH1=(H2(I-1)+H1(I-1))/2.
          QQ2=(Q1(I+1)+Q1(I))/2.
          QQ1=(Q1(I)+Q1(I-1))/2.
      8   Q2(I)=(QQ2+QQ1)/2,-DT*(QQ2**2/HH2-QQ1**2/
      1   HH1)/DX+9.81*HH*(HH2-HH1)/DX+9.81*HH*SF)
          Q2(IMAX)=(H2(IMAX1)+H1(IMAX1)-2.*HO)/2.
      1   *SQRT(9.81*(H2(IMAX1)+H1(IMAX1))/2.)
          DO 9 I=1,IMAX
      9   Q1(I)=Q2(I)
          DO 11 I=1,IMAX1
     11   H1(I)=H2(I)
          IF(N/10*10.LT.N) GO TO 100
          WRITE(6,12) T
     12   FORMAT(F10.3)
          WRITE(6,13)(Q2(I),I=1,IMAX)
          WRITE(6,13)(H2(I),I=1,IMAX1)
     13   FORMAT(15F8.3)
          IF(N.LT.2000) GO TO 100
          STOP
          END
```

Description of variables:
    Q1(I), Q2(I)   old and new discharge values
    H1(I), H2(I)   old and new water depth values
    QQ(I)   upstream flood hydrograph
    IMAX   the number of cross sections
    CMAN   the Manning K coefficient
    HO   the initial water depth
    DT   time step
    DX   space step
    DDT   time step in reading QQ values

Data values: IMAX=30, DT=10 s, DX=1000 m, CMAN=30 $m^{1/3}$/s, HO=1 m, DDT=3600 s, QQ=according to Fig. 4.3.)

PROGRAM FOR IMPLICIT INTEGRATION

```
C     IMPLICIT FD SOLUTION OF FLOOD PROPAGATION
C     IN OPEN CHANNEL
      DIMENSION Q1(50), Q2(50),H1(50),H2(50),QQ(50)
      READ(5,5) DT,DDT,DX,HO,CMAN,IMAX
    5 FORMAT(5F7.0,I4)
      IMAX1=IMAX-1
      DO 10 I=1,IMAX
```

```
      Q1(I)=0.
      Q2(I)=0.
      H1(I)=HO
   10 H2(I)=HO
      READ(5,6)(QQ(N),N=1,21)
    6 FORMAT(10F7.0)
      T=0.
      N=0
  100 T=T+DT
      N=N+1
      DO 20 I=1,IMAX1
   20 H2(I)=H1(I)-DT/DX*(Q1(I+1)-Q1(I))
      L=INT(T/DDT)+1
      Q2(1)=QQ(L)+(QQ(L+1)-QQ(L))*(T-(L-1)*DDT)/DDT
      Q2(IMAX)=(H2(IMAX1)-HO)*SQRT(9.81*H2(IMAX1))
      ITER=0
   30 ITER=ITER+1
      IF(ITER.GT.100) STOP 1
      DIFMX=0.
      DO 40 I=2,IMAX1
      QQ1=(Q2(I)+Q2(I-1)+Q1(I)+Q1(I-1))**2./16./
     1  H2(I-1)
      QQ2=(Q2(I+1)+Q1(I+1)+Q2(I)+Q1(I))**2/16./
     1  H2(I)
      SF=((Q2(I)+Q1(I))/2./CMAN)**2/((H2(I)+
     1  H2(I-1))/2.)**3.333
      TEMP=Q2(I)
      Q2(I)=Q1(I)-DT*(QQ2-QQ1)/DX-9.81*DT/2./DX*
     1  (H2(I)**2-H2(I-1)**2)-9.81*DT*(H2(I)+
     1  H2(I-1))/2.*SF
      DIF=ABS(TEMP-Q2(I))
      IF(DIF.GT.DIFMX) DIFMX=DIF
   40 CONTINUE
      IF(DIFMX.GT..001) GO TO 30
      DO 60 I=1,IMAX
      Q1(I)=Q2(I)
   60 H1(I)=H2(I)
      IF(N/10*10.LT.N) GO TO 100
      WRITE(6,65) T
   65 FORMAT (F10.3)
      WRITE(6,70)(Q1(I), I=1,IMAX)
      WRITE(6,70)(H1(I), I=1,IMAX1)
   70 FORMAT(15F8.2)
      IF(N.LT.2000) GO TO 100
      STOP
      END
```

The variables and data values are described above. The solution of the deriving algebraic system with respect to Q2(1, ..., IMAX) is performed ny successive iterations. The convergence test is equal to 0.001 m³/s.

74 OPEN CHANNEL FLOW

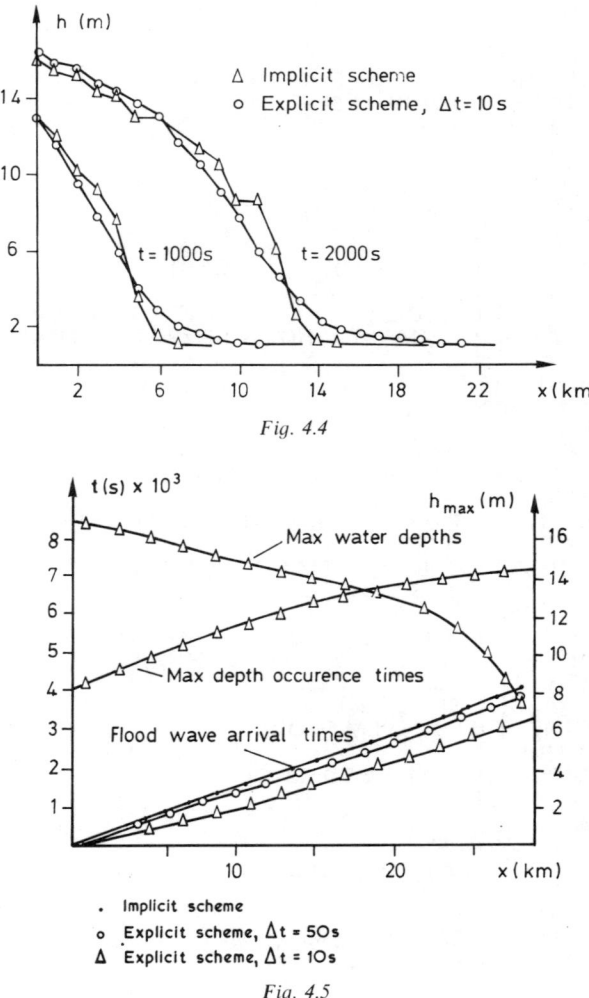

Fig. 4.4

Fig. 4.5

Figure 4.4 gives the water profile for 1000 s and 2000 s after the start of the flow, computed by the two methods.

Figure 4.5 illustrates the flood wave arrival times, the times of maximum water depth and the maximum water depth along the channel. From Figs. 4.4 and 4.5 it can be concluded that the implicit scheme does not contain much inherent numerical diffusion but as a fully centred scheme contains some numerical dispersion, while the explicit scheme implies strong numerical diffusion proportional to the time step Δt. Numerical diffusion is distinguishable most at the arrival times (it implies a faster propagation of the flood front) and less apparent in the maximum water depth values and their occurrence times.

If the mathematical model consisting of Equations 4.1 and 4.2 were to be applied, and the velocity values were unknown on the boundaries (case of given h or given discharge) the application of the above finite differences schemes would face among others the following difficulties:

(1) A fictitious value V_0 outside the flow domain would be required for the computation of the unknown V_1.

(2) In case of given $Q(=Vh)$, the V_1^{n+1} value would have to be computed from a $h_1^{n+\frac{1}{2}}$ value at a different time.

The above difficulties together with others more basic (such as the description of the propagation of shock waves) can be resolved only through the use of the method of characteristics.

The method of characterwistic which has a general applicability will be applied here to the simple case of flow in a rectangular channel. The form of Equations 4.1 and 4.2 in this case becomes:

$$\frac{\partial h}{\partial t} + h\frac{\partial V}{\partial x} + V\frac{\partial h}{\partial x} = 0 \tag{4.25}$$

$$\frac{\partial V}{\partial t} + V\frac{\partial V}{\partial x} + g\frac{\partial h}{\partial x} = g(S_0 - S_f) \tag{4.26}$$

The multiplication of Equation 4.25 by g, the substitution of gh by c^2 and the subsequent addition and subtraction of Equations 4.25 and 4.26 lead to the equations:

$$\frac{\partial}{\partial t}(V \pm 2c) + (V \pm c)\frac{\partial}{\partial x}(V \pm 2c) = g(S_0 - S_f) \tag{4.27}$$

which after the substitution $(dx/dt) = V \pm c$ can be written as:

$$\frac{d}{dt}(V \pm 2c) = g(S_0 - S_f) \tag{4.28}$$

valid on the characteristic curves,

$$\frac{dx}{dt} = V \pm c \tag{4.29}$$

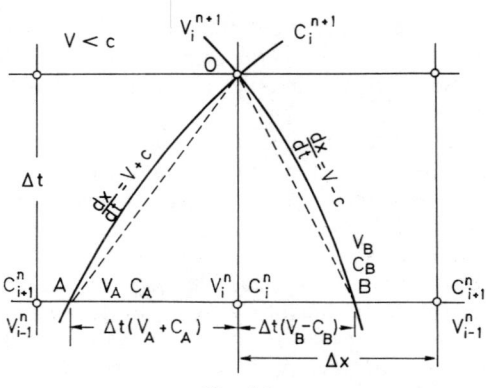

Fig. 4.6

Two characteristic curves, $(dx/dt) = V+c$ and $(dx/dt) = V-c$ pass from each point of the x–t space. Along these curves the quantities $V \pm 2c$ vary at a rate analogous to the local values of $g(S_0 - S_f)$. S_f can be approximated using Equation 4.3.

Various schemes for the numerical integration of Equations 4.28 and 4.29, explicit or implicit, can be found in the recent literature[1,2]. The integration to be presented here is explicit with respect to time and is performed not on the characteristic curves themselves but on an orthogonal grid in x–t space. The explicit character of the integration requires the satisfaction of the Courant stability criterion, imposing a relation between Δt and Δx,

$$(V+c)\Delta t \approx c\Delta t \leqslant \Delta x \qquad (4.30)$$

If Equation 4.30 is satisfied, then the two characteristics passing from point 0 of Fig. 4.6, for a small Δt, can be approximated by straight lines of slopes $(dx/dt) = V \pm c$, where V and c are approximately constant. These lines intersect the previous time level t_n at points A and B. The explicit integration of Equation 4.29 permits the approximation

$$V_i^{n+1} + 2c_i^{n+1} = V_A^n + 2c_A^n + \Delta t g(S_{0A} - S_{fA}^n) \qquad (4.31)$$

$$V_i^{n+1} - 2c_i^n = V_b^n - 2c_B^n + \Delta t g(S_{0A} - S_{fA}^n) \qquad (4.32)$$

The V_A, C_A, V_B, C_B values can be calculated through a simple linear interpolation from V_{i-1}^n, C_{i-1}^n, V_i^n, C_i^n, V_{i+1}^n, C_{i+1}^n known values. The location of points A and B are calculated from the integration of Equation 4.28,

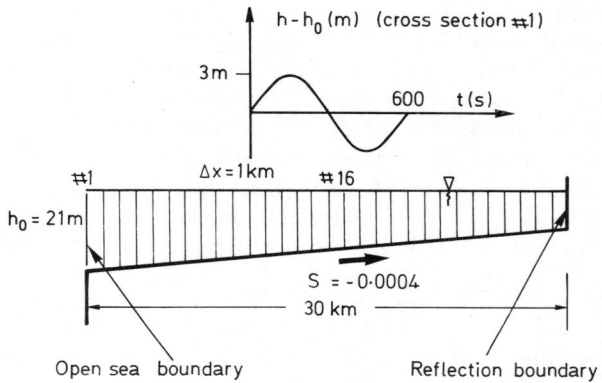

Fig. 4.7

$$\frac{\Delta x_L}{\Delta t} = V_A + c_A \qquad (4.33)$$

$$\frac{\Delta x_R}{\Delta t} = V_B - c_B \qquad (4.34)$$

The procedure for the computation of V_A, C_A, V_B, C_B and subsequently of V_i^{n+1}, C_i^{n+1}, through the application of Equations 4.31 and 4.32 is programmed in the following example.

EXAMPLE 4.2

The propagation of a long wave of period $T = 600$ s along an estuary of rectangular cross section, of bottom slope 0.4% and a friction coefficient $K = 30$ m$^{1/3}$/s is to be described. The amplitude of the tidal wave at the sea boundary is 3 m.

What are the water depth and water velocity variations at 15 km from the sea?

The flow field is schematised in Fig. 4.7. The initial conditions consist of a horizontal free surface and zero flow along the flow field. The upstream and downstream water depths are 9 m and 21 m respectively.

The boundary conditions adopted here are the reflection condition $V = 0$ on the upstream end and the given h fluctuation on the downstream end. For the computation of water depth at the upstream end, Equation 4.31 is applied with $V_i = 0$; for the computation of the velocity at the downstream boundary, Equation 4.32 is applied and solved for V_i^{n+1} with known $c_i^{n+1} = \sqrt{(gh_i^{n+1})}$. The use of the properties of characteristics permits the numerical treatment of complicated boundary conditions (relating the velocity and depth functions) in a simple and efficient way.

The maximum wave propagation speed appears at the downstream boundary, i.e. $c = \sqrt{21g} = 14.3$ m/s. If a space step $\Delta x = 1000$ m is adopted, implying 30 reaches along the estuary and 31 cross-sections ($i_{max} = 31$) then the maximum allowable time step is $\Delta t = 1000/14.3 = 69.5$ s.

In order to control the numerical diffusion, superimposed on the natural frictional decay processes, a $\Delta t = 60$ s value is used.

The computational procedure described by Equations 4.31 and 4.32 is programmed in FORTRAN as follows:

```
C      LONG WAVES IN OPEN CHANNEL METHOD OF
C      CHARACTERISTICS
       DIMENSION U(50),Y(50),C(50),UA(50),UB(50),
     1  CA(50),CB(50),SFA(50),SFB(50)
       READ(5,5) IMAX,DT,DX,SLOPE,PER,AMPL,FR
  5    FORMAT(I4,6F7.0)
       IMAX1 = IMAX − 1
       READ(5,10)(Y(I),I = 1,IMAX)
 10    FORMAT(10F7.0)
       DO 15 I = 1,IMAX
       U(I) = 0.
       UA(I) = 0.
       UB(I) = 0.
       SFA(I) = 0.
       SFB(I) = 0.
 15    C(I) = SQRT(9.81*Y(I))
       YO = Y(1)
       T = 0.
       N = 0
100    T = T + DT
       N = N + 1
       Y(1) = YO + AMPL*SIN(6.2832*T/PER)
       C(1) = SQRT(9.81*Y(1))
       DO 20 I = 1,IMAX
       IF(I.EQ.1) GO TO 21
       A = (U(I−1) − U(I))*DT/DX
       B = (C(I−1) − C(I))*DT/DX
       UA(I) = (U(I) + A/(1. − B)*C(I))/(1. − A − A*B/(1. − B))
       CA(I) = (C(I) + B/(1. − A)*U(I))/(1. − B − A*B/(1. − A))
       YA = CA(I)**2/9.81
       SFA(I) = UA(I)*ABS(UA(I))/FR**2/YA**1.333
 21    CONTINUE
       IF(I.EQ.IMAX) GO TO 22
       G = (U(I+1) − U(I))*DT/DX
       D = (C(I+1) − C(I))*DT/DX
```

```
      UB(I) = (U(I) + G*C(I)/(1.-D))/(1.+G+G*D/(1.-D))
      CB(I) = (C(I) - D*U(I)/(1.+G))/(1.-D+G*D/(1.+G))
      YB = CB(I)**2/9.81
      SFB(I) = UB(I)*ABS(UB(I))/FR**2/YB**1.333
22    CONTINUE
20    CONTINUE
      DO 25 I=2,IMAX1
      U(I) = (DT 9.81*(2.*SLOPE - SFA(I) - SFB(I)) +
     1 UA(I) + UB(I) + 2.*CA(I) - 2.*CB(I))/2.
      C(I) = (DT*9.81*(SFB(I) - SFA(I)) + UA(I) -
     1 UB(I) + 2.*CA(I) + 2.*CB(I))/4.
      Y(I) = C(I)**2/9.81
25    CONTINUE
      U(1) = DT*9.81*(SLOPE - SFB(1)) + 2.*C(1) +
     1 UB(1) - 2.*CB(1)
      C(IMAX) = (DT*9.81*(SLOPE - SFA(IMAX)) + UA(IMAX) +
     1 2.*CA(IMAX))/2.
      Y(IMAX) = C(IMAX)**2/9.81
      WRITE(6,30) T
30    FORMAT(5X, 'TIME', F10.2)
      WRITE(6,35)(Y(I), I=1,IMAX)
35    FORMAT(10F10.4)
      WRITE(6,35) (U(I), I=1,IMAX)
      IF(N.LT.500) GO TO 100
      STOP
      END
```

Description of variables:
  $U(I)$ = velocity values
  $Y(I)$ = water depth values
  $C(I)$ = speed of propagation
  $UA(I), UB(I)$ = past velocity values
  $CA(I), CB(I)$ = past speed of propagation values
  $SFA(I), SFB(I)$ = past energy slope values
  SLOPE = bottom slope
  PER = period of tidal wave
  AMPL = amplitude of tidal wave
  FR = friction coefficient

The application is done for values of the input parameters IMAX = 31, DT = 60 s, DX = 1000 m, SLOPE = -0.0004, PER = 600 s, AMPL = 3 m and FR = 30 m$^{1/3}$/s. The water depth and velocity occurring at a distance 15 km from the sea is illustrated in Fig. 4.8.

The establishment of periodicity after the first few cycles is detected. Figure 4.9 contains the wave height variation along the estuary.

It is left to the reader to check experimentally (through various combinations of $\Delta t, \Delta x$) the intrusion and influence of numerical diffusion in the process of continuous decay of the wave height.

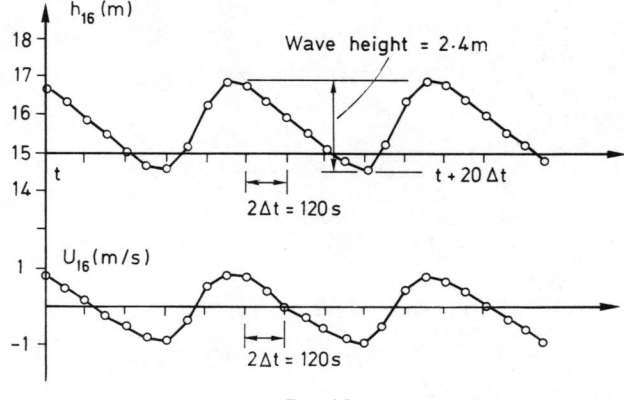

Fig. 4.8

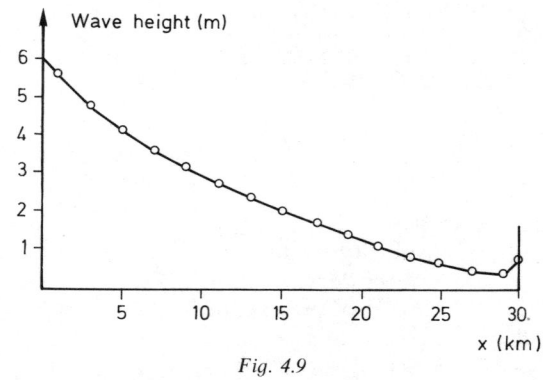

Fig. 4.9

4.3. STEADY NON-UNIFORM FLOW. BACKWATER CURVE ANALYSIS

4.3.1. Mathematical formulation of steady non-uniform flow

In most hydraulic problems it is necessary to compute the variation of the water profile along open channels under conditions of steady but spatially varying (non-uniform) flow. The V (velocity) and h (water depth) magnitudes are functions of space (x). In the case of smooth variations of V and h along a channel portion much longer than the water depth, the flow is characterised as gradually varying, otherwise as rapidly varying.

The mathematical model of steady, gradually varying flow derives

OPEN CHANNEL FLOW 81

from Equations 4.1 and 4.2 after dropping the time derivatives. Thus the continuity and conservation equations take the form,

$$Q = AV = \text{constant} \tag{4.35}$$

$$\frac{V}{g}\frac{dV}{dx} + \frac{dh}{dx} = S_0 - S_f \tag{4.36}$$

The non-linear term can undergo successive transformations as follows:

$$\frac{V}{g}\frac{dV}{dx} = \frac{1}{2g}\frac{dV^2}{dx} = \frac{1}{2g}\frac{d}{dx}(Q^2/A^2) = -\frac{2Q^2}{2g}\frac{dh}{dx}\frac{B}{A^3} \tag{4.37}$$

and the final form of Equation 4.36 solved with respect to the water depth gradient becomes:

$$\frac{dh}{dx} = \frac{S_0 - S_f}{1 - \frac{Q^2 B}{gA^3}} \tag{4.38}$$

The energy line slope S_f is also calculated here by the formulae for uniform flow with values of the local flow magnitudes. The normal flow depth h_n, i.e. the flow depth for uniform flow with given Q and S_0 values, is calculated through Equation 4.39(a),

$$Q = KA(h)R(h)^{2/3}S_0^{1/2} \tag{4.39(a)}$$

and the critical flow situation is defined through Equation 4.39(b),

$$Q^2 B/gA^3 = 1 \tag{4.39(b)}$$

which can be solved to give the critical depth (h_c) and slope (S_{ocr}).
The following cases can be distinguished:

$$h > h_n \rightarrow S_f < S_0 \tag{4.40(a)}$$

$$h < h_n \rightarrow S_f > S_0 \tag{4.40(b)}$$

The expression $Q^2 B/gA^3$ can be either greater or less than unity for supercritical (torrential) and subcritical (fluvial) flow situations, respectively. Investigation of the sign of dh/dx in Equation 4.38, according to the sign of the numerator and denominator of the r.h.s.

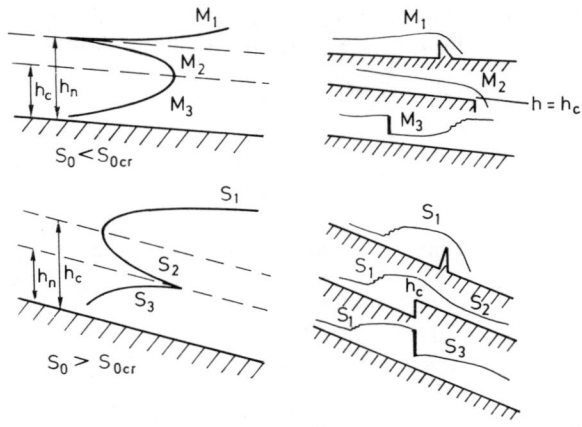

Fig. 4.10

of Equation 4.38, leads to the following classification:
Channel slope S_0 greater than the critical S_{ocr}, i.e. $h_n < h_c$:

$$h > h_c > h_n, \quad dh/dx = +, \quad \text{Curve } S_1$$

$$h_c > h > h_n, \quad dh/dx = -, \quad \text{Curve } S_2$$

$$h_c > h_n > h, \quad dh/dx = +, \quad \text{Curve } S_3$$

Channel slope S_0 less than the critical S_{ocr}, i.e. $h_n > h_c$:

$$h > h_n > h_c, \quad dh/dx = +, \quad \text{Curve } M_1$$

$$h_n > h > h_c, \quad dh/dx = -, \quad \text{Curve } M_2$$

$$h_n > h_c > h, \quad dh/dx = +, \quad \text{Curve } M_3$$

These are some basic backwater curves and their form together with examples of their occurrence are illustrated in Fig. 4.10.

The integration of Equation 4.38 is a problem of the numerical solution of an ordinary differential equation with a given boundary condition. The known boundary condition is the water depth h at a 'control section' (upstream in the case of $h < h_c$, downstream in the case of $h > h_c$).

4.3.2. Numerical solution of steady non-uniform flow equations

Equation 4.38 is a non-linear ordinary differential equation as S_f, B and A are non-linear functions of h. Several methods for its numerical integration are included in the recent bibliography[7] on the numerical solution of ordinary differential equations.

Here, two simple numerical procedures will be illustrated, the integration methods for (1) variable and (2) constant space steps.

(1) Variable step method

This is a very simple method applicable only in the case of prismatic channels of constant cross-section. According to this method, Equation 4.36 is rewritten in the form

$$\frac{\Delta}{\Delta x}\left(\frac{V^2}{2g}\right) + \frac{\Delta h}{\Delta x} = S_0 - \bar{S}_f \quad (4.41)$$

where $\bar{S}_f$ is the mean energy slope value along a reach of length Δx. Equation 4.41 is solved for Δx with predetermined Δh,

$$\Delta x = \frac{\Delta\left(h + \dfrac{V^2}{2g}\right)}{S_0 - \bar{S}_f} = \frac{\dfrac{h_{i+1} + V_{i+1}^2}{2g} - \dfrac{h_i + V_i^2}{2g}}{S_0 - \bar{S}_f} \quad (4.42)$$

The distance Δx between two sections with known flow magnitudes h_{i+1}, V_{i+1}, h_i, V_i is computed by the aid of Equation 4.42. The mean value $\bar{S}_f$ can be computed from the mean flow values $\bar{V} = (V_i + V_{i+1})/2$ and $\bar{h} = (h_i + h_{i+1})/2$. Starting from the control section (h_1, V_1) the distances between the sections $1 \div 2$, $2 \div 3 \ldots n-1 \div n$ are computed. An application of the method of variable steps is given in the following example.

EXAMPLE 4.3

What is the length along which the flow regime is influenced upstream of a dam built across a rectangular channel of width 5 m, bottom slope $S_0 = 1\%$, Manning coefficient $K = 50 \text{ m}^{1/3}/\text{s}$ and discharge $Q = 55.4 \text{ m}^3/\text{s}$ if the water depth at the dam is 8 m?

The normal depth h_n is computed by means of Equation 4.39(a) to give $h_n = 5$ m, and the critical depth is computed by means of a

84 OPEN CHANNEL FLOW

relation deriving from Equation 4.39(b) for rectangular cross-sections,

$$h_c = \sqrt[3]{\left(\frac{Q^2}{gB^2}\right)} = 2.3 \text{ m} \qquad (4.43)$$

The normal flow is subcritical, $h_n > h_c$. The critical slope is found from Equation 4.39(b) to be $S_{ocr} = 7\%$, an M_1 backwater curve.

For the computation of $\Delta x_1 \ldots \Delta x_n$, the flow depths from $h_1 = 8$ m to $h_n = 5$ m are discretised into steps $\Delta h = 0.1$ m, and thus 30 reaches are formed. The computation of the Δx_i for these 30 reaches is programmed as follows:

```
C      BACKWATER CURVES VARIABLE STEP METHOD
       DIMENSION H(50), V(50), DX(50)
       READ(5,1) CMAN,Q,SO,IMAX
1      FORMAT(3F7.0,I4)
       READ(5,2) H(1),H(IMAX)
2      FORMAT(2F7.0)
       IMAX1=IMAX-1
       DO 3 I=1,IMAX
       H(I)=H(1)+(H(IMAX)-H(1))*(I-1)/IMAX1
3      V(I)=Q/H(I)/B
       DO 4 I=1,IMAX1
       HH=(H(I)+H(I+1))/2.
       A=B*HH
       P=2.*HH+B
       R=A/P
       SF=Q**2/CMAN**2/R**1.333/A**2
4      DX(I)=(H(I+1)+V(I+1)**2/2./9.81-H(I)-
      1 V(I)**2/2./9.81)/(SO-SF)
       SUM=0.
       DO 5 I=1,IMAX1
5      SUM=SUM+DX(I)
       WRITE(6,6) SUM
6      FORMAT(F10.2)
       WRITE(6,7)(DX(I),I=1,IMAX1)
7      FORMAT (10F10.2)
       STOP
       END
```

Description of variables:
   H(I)   = water depth
   V(I)   = water velocity
   DX(I) = length of reaches
   CMAN = Manning's K coefficient
   Q   = the discharge
   SO   = the bottom slope

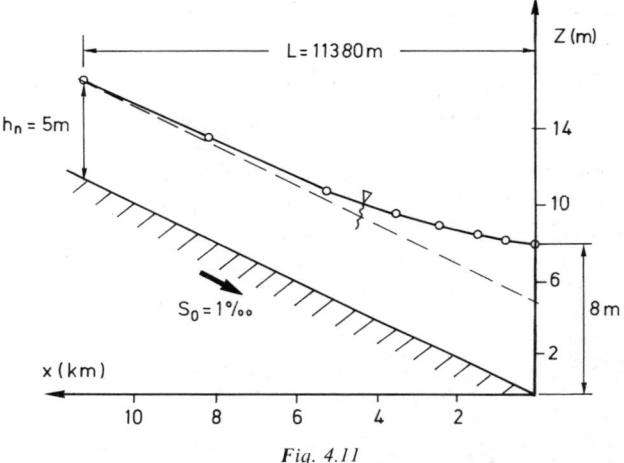

Fig. 4.11

For the given input values, CMAN = 50 m$^{1/3}$/s, Q = 55.4 m^3/s and SO = −0.001, the application of the program leads to the M_1 curve described in Fig. 4.11. The total length L ($L = \sum_{i=1}^{30} \Delta x_i$) is found to be 11.38 km.

(2) Constant step method

The previous method is not applicable to non-prismatic channels. In the case of a non-prismatic channel Equation 4.42 is solved with respect to Δh_s (difference in specific energy).

$$\Delta h_s = \Delta(h + V^2/2g) = \Delta x(S_0 - \bar{S}_f) \tag{4.44}$$

where Δ is the known difference operator.

The calculation of h_{i+1}, V_{i+1} from V_i, h_i is performed by successive iterations. Starting with a high value for h_{i+1} (forming the upper limit of the expected h values along the channel) Equation 4.44 is applied and $h_{s(i+1)}$ is computed. The same magnitude is computed from the relation defining h_s,

$$h_s = h + \frac{V^2}{2g} = h + \frac{Q^2}{2g(bh)^2} \tag{4.45}$$

The two $h_{s(i+1)}$ values are compared and h_{i+1} is decreased in a regular or accelerated form down to the coincidence of h_s values computed

86 OPEN CHANNEL FLOW

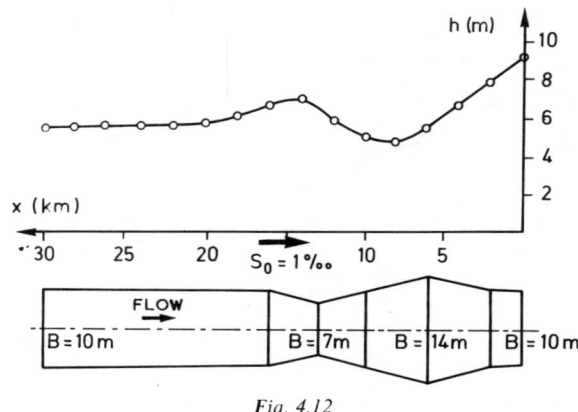

Fig. 4.12

from Equations 4.44 and 4.45. The method is actually a form of numerical computation of a solution (a zero) of a complicated algebraic equation. Of course the choice of the first approximation of h_{i+1} and the variation step Δh_{i+1} are dominant factors for the computational time. Methods for fast convergence to this solution, from an arbitrary starting point, may be found in Hamming[11].

EXAMPLE 4.4

In a channel of rectangular cross-section with varying width, as given in Fig. 4.12, with $Q = 100$ m³/s, $S_0 = 1\%$ and $K = 30$ m$^{1/3}$/s, a backwater curve is caused by a dam of height 7 m.

The normal flow depth for width $B = 10$ m and $S_0 = 1\%$ is found to be $h_n = 5.6$ m and the corresponding $h_c = 2.17$ m.

The flow over the dam is known to occur with critical water depth, so the depth at the downstream control section is $7 + 2.17 = 9.17$ m. This varies up to an upstream uniform depth of 5.6 m. The method of constant steps is applied after the discretisation of the flow domain (taken equal to 30 km) by means of a spatial step $\Delta x = 1$ km. The water depth and velocity magnitudes are computed on each cross-section using Equation 4.44. The procedure is programmed as follows:

```
C     BACKWATER CURVES FIXED STEP METHOD
      DIMENSION H(100),V(100),B(100),HS(100),HST(100)
      READ (5,1) IMAX, DX, CMAN,Q,SO, H(1)
1     FORMAT (I4,5F7.0)
      READ (5,2) (B(I),I=1,IMAX)
```

```
      2   FORMAT (10F7.0)
          V(1)=Q/B(1) / H(1)
          HS(1)=V(1)**2/2./9.81+H(1)
          DO 3 I=2,IMAX
          H(I)=10.
          DIFN=1.
          DH=1.
     20   DIF=DIFN
          NTEST=0
          H(I)=H(I)-DH
     15   V(I)=Q/B(I)/H(I)
          HH=(H(I-1)+H(I))/2.
          BB=(B(I-1)+B(I))/2.
          VV=Q/BB/HH
          R=HH*BB/(BB+2.*HH)
          SF=VV**2/CMAN**2/R**1.33
          HS(I)=HS(I-1)+(SO-SF)*DX
          HST(I)=H(I)+V(I)**2/2./9.81
          DIFN=HST(I)-HS(I)
          IF(NTEST.GT.O) GO TO 16
          IF(DIF*DIFN.GT.O.) GO TO 20
          H(I)=H(I)+DH
          NTEST=1
          GO TO 15
     16   DH=DH/10.
          IF(DH.GT..0001) GO TO 20
      3   CONTINUE
          WRITE (6,4)(I,H(I), V(I),I=1,IMAX
      4   FORMAT (I6,2F10.3)
          STOP
          END
```

Description of variables:

- H(I) = water depth
- V(I) = velocity
- B(I) = cross section width
- HS(I) }
- HST(I) } = specific head values computed through Equations 4.44 and 4.45 respectively
- DX = space step
- CMAN = Manning coefficient
- Q = discharge
- SO = bottom slope

For input values IMAX = 30, DX = -1000 m, CMAN = 30 m$^{1/3}$/s, Q = 100 m^3/s, SO = 0.001, H(1) = 5.6 m and cross-section widths B(I), as given in Fig. 4.12, the computed depth values are depicted in Fig. 4.12.

5
Groundwater flow

5.1. MATHEMATICAL MODELS FOR FLOW IN POROUS MEDIA

The mathematical models for flow in porous media are founded on Darcy's hypothesis linearly correlating the seepage velocity to the hydraulic head gradient,

$$\vec{V} = -K \, \text{grad} \, \varphi \tag{5.1}$$

where $\vec{V}$ is the seepage velocity, K the hydraulic conductivity or permeability and φ the hydraulic head (flow potential) measured from any reference datum.

The substitution of Equation 5.1 in the continuity equation, which in the case of incompressible flow has the form,

$$\nabla \cdot \vec{V} = 0 \tag{5.2}$$

leads to the general form of the field equation for flow in porous media

$$\nabla^2 \varphi = 0 \tag{5.3}$$

The boundary conditions completing Equation 6.3 can have the following forms:

(1) Impermeable boundary: $\partial \varphi / \partial n = 0$ where n denotes the direction perpendicular to the boundary.

(2) Reservoir boundary: φ is a given constant along the boundary.

(3) Seepage surface: φ varies linearly with height, $\varphi + kz = \text{const.}$, along the boundary.

(4) Free surface boundary: this is the most complicated boundary as its geometry is not predetermined. In unsteady flow situations the conditions holding on this boundary are:

 (i) Kinematic boundary condition

$$\frac{\partial h}{\partial t} = \frac{K}{n} \left(\frac{\partial \varphi}{\partial z} - \frac{\partial h}{\partial x} \frac{\partial \varphi}{\partial x} - \frac{\partial h}{\partial y} \frac{\partial \varphi}{\partial y} \right) \tag{5.4}$$

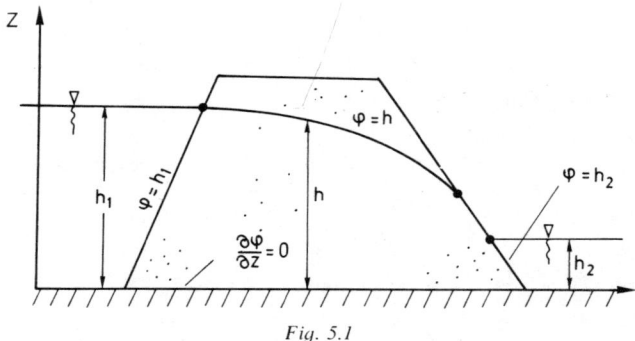

Fig. 5.1

(ii) Dynamic boundary condition

$$\varphi = h \qquad (5.5)$$

where h is the absolute elevation of the free surface, according to the notations of Fig. 5.1.

In the case of flow in an almost horizontal underground aquifer, with small flow depth and head gradient, the Dupuit approximation can be introduced (Remson et al.[21]). This approximation refers to the uniformity of the seepage velocity and flow potential φ along the aquifer depth. In this case the mathematical model for non-steady flow in the x, y directions in an aquifer contains as unknown function, the absolute elevation of the pressure surface (the flow hydraulic head). It can be represented by the Boussinesq equation and has the following form, according to the notation of Fig. 5.2.

$$S \frac{\partial h}{\partial t} = \frac{\partial}{\partial x}\left(K_x c \frac{\partial h}{\partial x}\right) + \frac{\partial}{\partial y}\left(K_y c \frac{\partial h}{\partial y}\right) + q \qquad (5.6)$$

Equation 5.6 is a general form applying both to unconfined flow (with free surface) and confined flow (between two impermeable layers). The medium may be anisotropic (i.e. $K_x \neq K_y$, variable in space).

The several parameters in Equation 5.6 have the following values:
(1) Confined flow: $S = \beta b$ where β is a coefficient describing the compressibility of the porous medium and the water, and $c = b$.
(2) Unconfined flow: $S = n$ (the porosity) and $c = d$.
The q term describes the local sinks or sources distributed along the aquifers (dimensions $L^3/T/L^2$).

In the case of confined flow in a homogeneous aquifer of constant

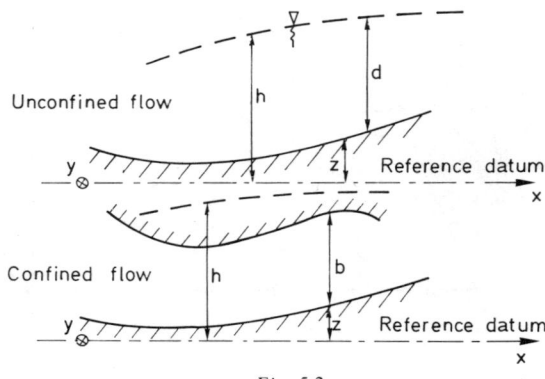

Fig. 5.2

thickness b, Equation 5.6 takes the form

$$\frac{\partial h}{\partial t} = \left(\frac{\partial^2 h}{\partial x^2} + \frac{\partial^2 h}{\partial y^2}\right)\frac{K}{\beta} \qquad (5.7)$$

In the case of an unconfined aquifer and flow over an horizontal impermeable layer, Equation 5.6 becomes,

$$\frac{\partial h}{\partial t} = \frac{K}{2n}\left(\frac{\partial^2 h^2}{\partial x^2} + \frac{\partial^2 h^2}{\partial y^2}\right) \qquad (5.8)$$

Equation 5.7 is the known parabolic equation while Equation 5.8 is a non-linear equation. It behaves basically as Equation 5.7 and in the case of small fluctuations of d around a mean value $\bar{d}$, it can be easily linearised to a parbolic form similar to Equation 5.7. The application of the mathematical models of flow in porous media (actually, special forms of parabolic and elliptic equations) will be demonstrated through the examples in Section 6.2.

5.2. APPLICATION OF MATHEMATICAL MODELS TO FLOW IN POROUS MEDIA

5.2.1. Numerical solution of the Darcy equation

EXAMPLE 5.1

A dam, with base 60 m wide and upstream water depth equal to 10 m, is built on a permeable ground layer, 80 m thick, with a permeability

GROUNDWATER FLOW 91

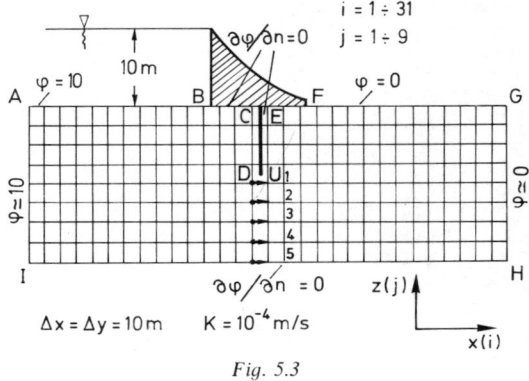

Fig. 5.3

coefficient, $K = 10^{-4}$ m/s. A sheetpile diaphragm extending halfway through the permeable soil reduces the seepage. The situation is illustrated in Fig. 5.3.

What is the seepage discharge under the dam?

Although the flow field extends an infinite distance upstream and downstream it is assumed here, for practical purposes, as the contribution of the far upstream reservoir portion to the seepage is rapidly reducing, that the flow field is bounded by the limits BC, CD, DE, EF and HI (impermeable boundaries where the normal velocity is suppressed) and the artificial limits AI and GH approximated to constant head lines 10 m and 0 m respectively.

For ease of computation, the sheetpile is not placed on a grid line but between two of them so that the φ function is not double valued at certain nodes.

The flow domain is discretised by means of a rectangular grid of mesh size $\Delta x = 10$ m, $\Delta y = 10$ m.

The Laplace equation holding for the flow potential is solved numerically and the hydraulic head is computed at the nodes of the established grid.

The numerical procedure is programmed as follows:

```
C     SEEPAGE UNDER A DAM FD SOLN OF LAPLACE EQN
      DIMENSION F(40,20)
      READ(5,1) IMAX,JMAX,DX,DY,FO
    1 FORMAT(2I4,3F7.0)
      DO 20 I=1, IMAX
      DO 20 J=1,JMAX
   20 F(I,J)=0.
      DO 2 I=1,13
    2 F(I,JMAX)=FO
      DO 3 I=19,31
```

```
    3   F(I,JMAX)=0.
        DO 4 J=1,8
        F(IMAX,J)=0.
    4   F(1,J)=10.
        IMAX1=IMAX-1
        JMAX1=JMAX-1
        ITER=0
  100   DIFMX=0.
        ITER=ITER+1
        IF(ITER.GT.1000) STOP 1
        DO 5 I=2,IMAX1
        DO 5 J=2,JMAX1
        FL=F(I-1,J)
        DXL=DX
        IF(I.EQ.16).AND.((J.EQ.6).OR.(J.EQ.7).OR.
       1  (J.EQ.8)) GO TO 7
        GO TO 8
    7   DXL=DX/2.
        FL=F(I,J)
    8   CONTINUE
        FR=F(I+1,J)
        DXR=DX
        IF(I.EQ.15).AND.((J.EQ.6).OR.(J.EQ.7).OR.
       1  (J.EQ.8))) GO TO 9
        GO TO 10
    9   DXR=DX/2.
        FR=F(I,J)
   10   CONTINUE
        DYO=DY
        FO=F(I,J+1)
        IF((J.EQ.8).AND.(I.LT.19).AND.(I.GT.12))
       1   GO TO 11
        GO TO 12
   11   FO=F(I,J)
   12   CONTINUE
        DYU=DY
        FU=F(I,J-1)
        IF(J.EQ.2) GO TO 13
        GO TO 14
   13   FU=F(I,J)
   14   CONTINUE
        DXX=DXR+DXL
        DYY=DYO+DYU
        TEMP=F(I,J)
        A=1./DXR/DXL+1./DYO/DYU
        F(I,J)=((FR/DXR+FL/DXL)/DXX+(FO/DYO+FU/DYU)/
       1  DYY)/A
        DIF=ABS(TEMP-F(I,J))
        IF(DIF.GT.DIFMX) DIFMX=DIF
    5   CONTINUE
        IF(DIFMX.GT..00001) GO TO 100
        DO 15 J=JMAX,1,-1
   15   WRITE(6,18)(F(I,J),I=1,IMAX)
   18   FORMAT(20F6.2)
        STOP
        END
```

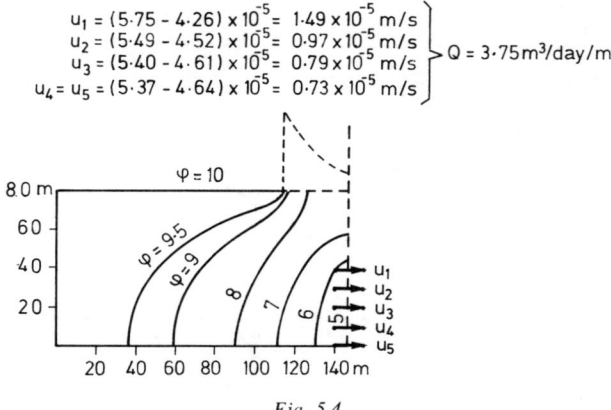

Fig. 5.4

Description of variables:
F(I,J) = head (potential) values along the flow field
DX,DY = space step in x, y directions
FO = upstream reservoir head
IMAX,JMAX = maximum number of nodes in the x, y directions

For data values IMAX=31, JMAX=9, DX=DY=10 m and FO=10 m, the head values are computed at the nodes of the flow field. The equipotential lines, drawn after interpolation, are shown in Fig. 5.4.

The numerical computation of the discharge under the dam is performed by computing the seepage velocities $u_1 \ldots u_5$ ($u = -K \cdot \partial \varphi / \partial x$) under the sheetpile. Numerical integration, by the trapezoidal rule, of the velocity profile over the distance between the impermeable bed and the lower end of the sheetpile gives the seepage discharge, $Q = 3.75$ m^3/day/m.

5.2.2. Numerical solution of the Boussinesq model

EXAMPLE 5.2

A porous stratum of length 1300 m, $K=1$ cm/s, and $n=0.4$ with a sloping impermeable bottom is laterally confined between two water masses. The initial water level is at 35 m.

A ditch is excavated at a distance 600 m from the left reservoir and a discharge of 500 m^3/day/m is pumped from the ditch with a simultaneous sudden drop of the level of the left reservoir from 35 m to 25 m (see Fig. 5.5).

94 GROUNDWATER FLOW

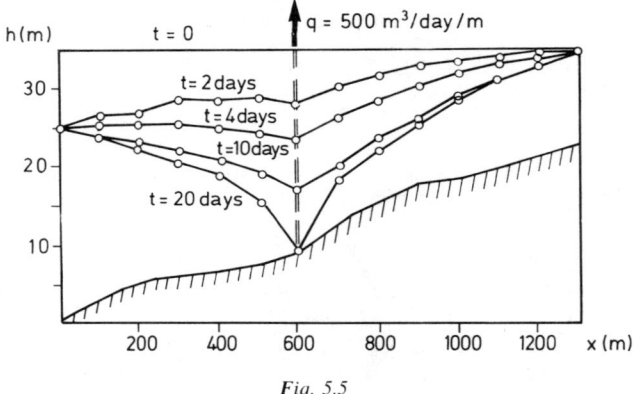

Fig. 5.5

The variation of the water level in the porous medium up to establishment of steady flow has to be computed.

The mathematical model applicable here is the Boussinesq model, as the horizontal extent of the aquifer is at least one order of magnitude greater than the water depth. In x, t space, for unconfined flow, its form derives from Equation 5.6.

$$\frac{\partial h}{\partial t} = \frac{K}{n} \frac{\partial}{\partial x}\left(d \frac{\partial h}{\partial x}\right) - q \tag{5.9}$$

where $d = h - z$, according to Fig. 5.2.

The numerical integration can be performed by means of the known explicit finite difference scheme for parabolic equations,

$$\frac{h_i^{n+1} - h_i^n}{\Delta t} = \frac{K}{n} \frac{[(d_{i+1}^n + d_i^n)(h_{i+1}^n - h_i^n) - (d_i^n + d_{i-1}^n)(h_i^n - h_{i-1}^n)]}{2\Delta x^2}$$

$$- q_i \tag{5.10}$$

Initial conditions consist of a constant h value ($h = 35$ m) and $q = 0$. The boundary conditions are: on the left reservoir boundary $h_{\text{left}} = 25$ m, on the right reservoir boundary $h_{\text{right}} = 35$ m. The discharge at the ditch is $q = -500/\Delta x$ m^3/day/m^2. The flow field discretisation is achieved through a spatial step $\Delta x = 100$ m. From the stability criterion for the explicit integration of parabolic equations and a mean value $\bar{d} = 20$ m, $K = 864$ m/day and $n = 0.4$, the critical time step is found to be

$$\Delta t \leqslant \frac{1}{2}\frac{\Delta x^2}{d}\frac{n}{K} \to \Delta t = 0.1 \text{ days} \qquad (5.11)$$

The numerical integration of Equation 5.10 is programmed in FORTRAN as follows:

```
C     1-DIMENSIONAL FLOW IN UNCONFINED AQUIFER
C     BOUSSINESQ EQN
      DIMENSION Z(50),H(50),D(50),HN(50),Q(50)
      READ(5,1) DX,DT,PERM,HO,POR,QO,IMAX
1     FORMAT (6F7.0I4)
      DO 5 I=1,IMAX
5     H(I)=HO
      READ(5,2) H(1)
2     FORMAT(F7.0)
      DO 6 I=1,IMAX
      Q(I)=0.
6     HN(I)=H(I)
      Q(7)=QO
      READ(5,3)(Z(I),I=1,IMAX)
3     FORMAT(10F7.0)
      IMAX1=IMAX-1
      T=0.
      N=0
100   T=T+DT
      N=N+1
      DO 8 K=1,IMAX
8     D(K)=H(K)-Z(K)
      DO 16 I=2,IMAX1
16    HN(I)=H(I)+DT*PERM/POR/DX**2/2.*((D(I+1)+
     1  D(I))*(H(I+1)-H(I))-(D(I)+D(I-1))*(H(I)-
     1  H(I-1)))-Q(I)*DT/POR
      DO 7 I=1,IMAX
7     H(I)=HN(I)
      WRITE(6,10) T
10    FORMAT(F10.2)
      WRITE(6,9)(H(I), I=1,IMAX)
9     FORMAT(10F10.3)
      IF(N.LT.300) GO TO 100
      STOP
      END
```

Description of variables:

- $Z(I)$ = impermeable bed elevation (as in Fig. 5.5)
- $H(I), HN(I)$ = old and new values of water elevation in the aquifer
- $D(I)$ = water depth in the aquifer
- $Q(I)$ = sink or source term (where it exists)
- PERM = permeability coefficient
- HO = initial water surface elevation

POR   = porosity coefficient
QO   = discharge pumped from 1 m length of ditch
Data values: $DX = 100$ m, $DT = 0.1$ days, $PERM = 864$ m/day, $HO = 35$ m, $POR = 0.4$, $QO = 5$ m^3/day/m^2, $IMAX = 14$, $H(1) = 25$ m.

The application of the program gives the evolution of the surface up to the time that steady flow conditions are reached, as illustrated in Fig. 5.5.

From Fig. 5.5 it can be seen that steady flow is established after 20 days. From the water depth at the ditch ($x = 600$ m) it seems that the discharge $q = 500$ m^3/day/m is the theoretical maximum that can be drawn according to Boussinesq's model. It should be noted that near the ditch and under heavy pumping conditions the sharp dip in the water surface and the appearance of a seepage surface on the ditch walls invalidate the Boussinesq model, and the maximum drawable discharge is seriously reduced.

EXAMPLE 5.3

A building is to be founded on an island with permeable soil. The island is assumed to be square, of side 260 m, with a horizontal base of impermeable soil. The surrounding water depth is 30 m.

The permeability coefficient is assumed equal to $K = 0.033$ m/day. For the foundation of the building, the water table is to be lowered from 30 m to 15 m, within the base of the building (a square of side 60 m, located centrally on the island). This is to be achieved by means of 4 well points positioned at the corners of the excavation site.

What is the discharge to be pumped out from each well?

A diagram of the construction site is given in Fig. 5.6.

From the symmetry of the flow field it is concluded that only the quarter plane OABC need be considered. In the case of steady unconfined flow in a 2-dimensional domain (x, y directions) and horizontal bed ($d = h$), Equation 5.6 takes the form,

$$\frac{K}{2}\left(\frac{\partial^2 h^2}{\partial x^2} + \frac{\partial^2 h^2}{\partial y^2}\right) + q = 0 \qquad (5.12)$$

Equation 5.12 is a Poisson type equation with respect to the function h^2. For its numerical integration, the well known centred finite difference schemes can be applied (Equation 2.37). The boundary conditions for this specific case become:
(1) Boundary lines AB, BC $h^2 = 900$ m^2.
(2) Boundary lines OA, OB $\partial h^2/\partial n = 0$.

GROUNDWATER FLOW 97

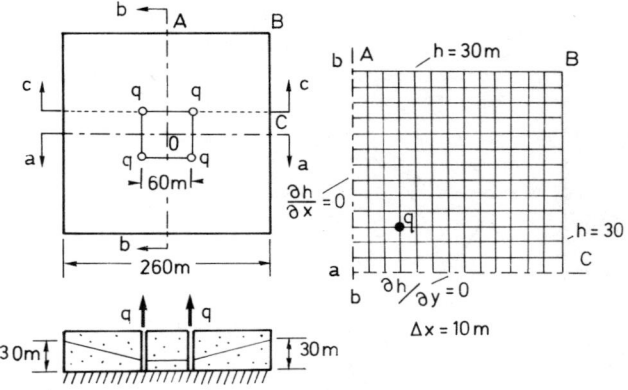

Fig. 5.6

The field is discretised through a square grid, with mesh size $DX = DY = 10$ m. The source or sink term, q, in Equation 5.12 is non-zero only at the node $i = 4$, $j = 4$.

The Poisson equation was solved for various q values, and through successive trials it was found that for a q value equal to $1 \text{ m}^3/\text{m}^2/\text{month}$ or $Q = 3.33 \text{ m}^3/\text{day}$ the water table in the foundation site does not exceed 15 m.

The computer program for this numerical solution is given below.

```
C     2-DIMENSIONAL FLOW IN UNCONFINED LAYER
C     BOUSSINESQ EQN
      DIMENSION H(20,20),Q(20,20)
      READ(5,1) DX,HO,QO,PERM,IMAX,JMAX
   1  FORMAT(4F7.02I4)
      IMAX1 = IMAX - 1
      JMAX1 = JMAX - 1
      DO 3 I = 1,IMAX
      DO 3 J = 1,JMAX
      Q(I,J) = 0.
   3  H(I,J) = HO
      Q(4,4) = QO
      ITER = 0
 100  DIFMX = 0.
      ITER = ITER + 1
      IF(ITER.GT.1000) STOP 1
      DO 4 I = 2,IMAX1
      DO 4 J = 2,JMAX1
      TEMP = H(I,J)
      H(I,J) = (H(I,J+1) + H(I,J-1) + H(I-1,J) + H(I+1,J))/4.
   1  - Q(I,J)/2./PERM*DX**2
      DIF = ABS(TEMP - H(I,J))
      IF(DIF.GT.DIFMX) DIFMX = DIF
```

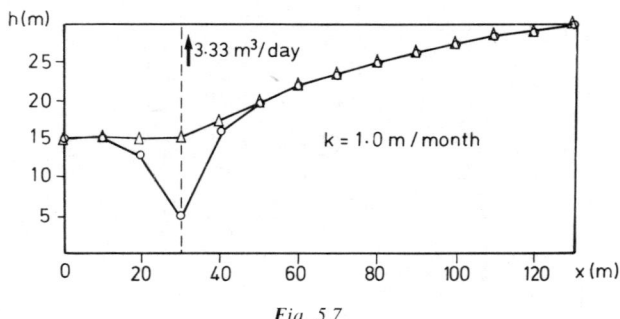

Fig. 5.7

```
4       CONTINUE
        DO 5 I=1,IMAX
5       H(I,1)=H(I,2)
        DO 6 J=1,JMAX
6       H(1,J)=H(2,J)
        IF(DIFMX.GT..00001) GO TO 100
        DO 7 I=1,IMAX
        DO 7 J=1,JMAX
7       H(I,J)=SQRT(H(I,J))
        DO 8 J=JMAX,1,-1
8       WRITE(6,9)(H(I,J),I=1,IMAX)
9       FORMAT(14F9.3)
        STOP
        END
```

Description of variables:
  H(I,J) = water table elevation in the medium
  Q(I,J) = source or sink discharge term
  DX   = spatial discretisation step
  HO   = initial water depth (around the island)
  QO   = the pumped discharge from each well point
  PERM   = permeability coefficient
  Data values: DX = 10 m, HO = 30 m, QO = 1 m^3/m^2/month, PERM = 1 m/month, IMAX = JMAX = 14.

Figure 5.7 contains characteristic profiles of the water table.

6
Advective diffusion and dispersion

6.1. MATHEMATICAL MODELS FOR DIFFUSION–DISPERSION OF MATTER

Problems of environmental pollution (for rivers, coasts and the atmosphere) can always be reduced to the solution of a mathematical model of diffusion–dispersion. The unknown quantity in these cases is the concentration, c, a scalar physical quantity, which can be the mass of a pollutant, or the salinity or temperature of the water, etc.

In the case of a pollutant mass it has to be distinguished if the quantity is conservative or not. A non-conservative quantity is one that has a continuously decaying mass due to biological or chemical reactions, even in absence of transport or diffusion. Such a non-conservative quantity can be a concentration of bacteria following an exponential decay with time. The principle of conservation of mass for a conservative quantity diluted or suspended in a moving fluid can be expressed mathematically through the quation:

$$\frac{dc}{dt} = \frac{\partial c}{\partial t} + \frac{\partial cu}{\partial x} = \frac{\partial F}{\partial x} \qquad (6.1)$$

where c is the concentration of the examined substance, u the fluid velocity in the x direction (it is assumed that the substance particles have the velocity of the surrounding fluid) and F is the mass of volume flux of the substance due to molecular diffusion. According to Fick's law this flux is expressed by,

$$F = \varepsilon \frac{\partial c}{\partial x} \qquad (6.2)$$

where ε is the molecular diffusion coefficient (usually a constant).

The concentration c is expressed as the ratio of the volume of a substance to the volume of the mixture or as a mass ratio. Units are usually expressed as ppt, ppm ... (parts per thousand, million, etc.).

The advective molecular diffusion in one dimension is thus decribed by the equation

$$\frac{\partial c}{\partial t} + \frac{\partial uc}{\partial x} = \varepsilon \frac{\partial^2 c}{\partial x^2} \tag{6.3}$$

which in the case of steady flow ($u = $ const.) becomes

$$\frac{\partial c}{\partial t} + u \frac{\partial c}{\partial x} = \varepsilon \frac{\partial^2 c}{\partial x^2} \tag{6.4}$$

The model can be easily generalised for a 3-D phenomenon.

In the case of a non-conservative quantity a sink term has to be added to the mass conservation equation. In the most usual case of exponential decay, due to a biological or other process, ($c = c_0 \, e^{-\lambda t}$) the r.h.s. of Equation 6.3 is completed by the addition of a term, $-\lambda c$. In the case of turbulent flow, as in most geophysical flows, the molecular diffusion term becomes negligible. The substitution of the velocity, u, and concentration, c, functions by their turbulent mean values $\bar{u}, \bar{c}$ and their turbulent fluctuations $u', c', (u = \bar{u} + u', c = \bar{c} + c')$ leads from Equation 6.4 to an equation of similar form,

$$\frac{\partial \bar{c}}{\partial t} + \bar{u} \frac{\partial \bar{c}}{\partial x} = \frac{\partial}{\partial x}\left(d \frac{\partial \bar{c}}{\partial x}\right) \tag{6.5}$$

where d is the turbulent diffusion coefficient (or eddy diffusion coefficient), generally a function of space and time.

Advective diffusion in closed or open conduits of small cross-section in comparison to their length permits the assumption that, at a section some distance from the pollutant source, the concentration distribution is almost uniform. The averaging of the diffusion equation over a cross-section correlates the sectional mean velocity V and concentration C through the dispersion equation

$$\frac{\partial (SC)}{\partial t} + \frac{\partial}{\partial x}(SVC) = \frac{\partial}{\partial x}\left(SD \frac{\partial C}{\partial x}\right) \tag{6.6}$$

where S is the section area and D the dispersion coefficient. The use of the mass continuity equation

$$\frac{\partial S}{\partial t} + \frac{\partial}{\partial x}(SV) = 0 \tag{6.7}$$

leads to further simplification of the dispersion equation

$$\frac{\partial C}{\partial t} + V \frac{\partial C}{\partial x} = \frac{\partial}{\partial x}\left(D \frac{\partial C}{\partial x}\right) \tag{6.8}$$

The boundary conditions completing the diffusion model of Equations 6.3 and 6.5 or the dispersion Equation 6.7 usually have the following forms:

(1) Reflection or wall boundary. The normal flux is zero, i.e. $F_n = \partial c/\partial n = 0$.

(2) Given concentration c or given flux $\partial c/\partial x$ boundary. These delineate the pollution source areas.

(3) Free transmission with no reflection boundary. A simple but efficient relation usable on this boundary is, $\partial^2 c/\partial n^2 = 0$.

6.2. NUMERICAL SOLUTION OF MATHEMATICAL MODELS OF DIFFUSION AND DISPERSION

From Equations 6.3, 6.5 and 6.8 it is evident that the diffusion and dispersion models are similar from a mathematical point of view and that they consist of mixed hyperbolic and parabolic differential operators. The hyperbolic part describes the advective transport of c and the parabolic part its diffusive transport due to molecular or turbulent processes. Consequently, the numerical integration of these models does not call for any further knowledge or techniques other than that used, for example, in the non-steady flow in open channels.

The use of an explicit scheme for the integration of Equation 6.4 requires the simultaneous satisfaction of two stability criteria,

$$c \frac{\Delta t}{\Delta x} \leqslant 1 \tag{6.9}$$

$$\varepsilon \frac{\Delta t}{\Delta x^2} \leqslant \tfrac{1}{2} \tag{6.10}$$

For fixed Δx, the Δt to be used is the smaller found from Equations 6.9 and 6.10. The basic problem in the integration of Equation 6.4 is the accurate description of the transport implied by u and the diffusion implied by ε.

The use of a finite difference scheme containing a rate of numerical diffusion in excess of the diffusion implied by the diffusion coefficients

102 ADVECTIVE DIFFUSION AND DISPERSION

Table 6.1

a	0	0.1	0.2	0.3	0.4	0.5	0.6	0.7
erf(a)	0	0.1125	0.223	0.328	0.428	0.520	0.604	0.678
a	0.8	0.9	1.0	1.1	1.2	1.3	1.4	1.5
erf(a)	0.742	0.797	0.843	0.880	0.910	0.934	0.952	0.966
a	1.6	1.7	1.8	1.9	2.0	2.1	2.2	2.3
erf(a)	0.976	0.983	0.989	0.993	0.995	0.997	0.998	0.999

ε (Equation 6.3) or d (Equation 6.5), or D (Equation 6.6) introduces errors to the numerical system.

The relation between the advective and diffusive transport is expressed by the Péclet number, a ratio appearing during the non-dimensionalisation of the terms in Equation 6.4,

$$\mathrm{Pe} = \frac{LU}{\varepsilon} \quad (6.11)$$

where L and U are the characteristic length and transport velocity of the phenomenon respectively. High Péclet numbers show that the advective transport process prevails over the diffusive one. Small Péclet numbers mean that diffusion (usually turbulent) dominates over the advection transport process.

Control of the accuracy of a numerical solution of the diffusion–dispersion model can be achieved by comparison between the numerical and an existing analytical solution. In the case of initial condition $c(x, 0) = 0$ and upstream boundary condition $c(0, t) = c_0$ the analytical solution of Equation 6.4 is

$$\frac{c}{c_0} = \tfrac{1}{2} \operatorname{erf} c\left(\frac{x-ut}{2(\varepsilon t)^{1/2}}\right) + \tfrac{1}{2} e^{\frac{ux}{\varepsilon}} \operatorname{erf} c\left(\frac{x+ut}{2(\varepsilon t)^{1/2}}\right) \quad (6.12)$$

where $\operatorname{erf} c = 1 - \operatorname{erf}$ and the values of erf(a), the error function, are given in Table 6.1.

The following example demonstrates the capability and some shortcomings of the numerical schemes commonly used for the description of an advective–diffusive process in relation to the values of the Péclet number.

EXAMPLE 6.1

A pollutant is diluted in an open channel with mean depth velocity $V = 1$ m/s. Its concentration at the source is constant, $c = 1$ ppt. The evolution of the mean depth concentration of the pollutant is to be numerically computed along the next few kilometers of the channel for values of dispersion coefficient $D = 1$, 0.1, 0.01 and 0.001 m^2/s.

Two numerical schemes are applied for the investigation of the dispersion process.
(1) Backward finite difference scheme. Equation 6.8 is approximated as follows:

$$\frac{c_i^{n+1} - c_i^n}{\Delta t} + V \frac{c_i^n - c_{i-1}^n}{\Delta x} = D \frac{c_{i+1}^n + c_{i-1}^n - 2c_i^n}{\Delta x^2} \tag{6.13}$$

(2) Fromm's scheme. Equation 6.8 is approximated, according to Section 2.3, Equation 2.35 as follows:

$$c_i^{n+1} = c_i^n + \frac{V \Delta t}{2\Delta t}(c_{i-1}^n - c_{i+1}^n + c_{i-2}^n - c_i^n) + \frac{V^2 \Delta t^2 + 4D\Delta t}{4\Delta x^2} \times$$

$$\times (c_{i+1}^n - 2c_i^n + c_{i-1}^n) + \frac{V^2 \Delta t^2 - 2V\Delta t \Delta x}{4\Delta x^2}(c_{i-2}^n - 2c_{i-1}^n + c_i^n)$$

$$\tag{6.14}$$

The FORTRAN programming is no more than a simple modification of previous similar programs and is left to the reader. (See, for example, programs in Section 4.2: 'Explicit FD Solution of Flood Propagation in Open Channel' and 'Long Waves in Open Channel, Method of Characteristics').

Some attention has to be paid to the application of Fromm's scheme. As the c_{i-2}^n values are used for the computation of c_i^{n+1}, the computation starts from c_3 assuming the c_1, c_2 are known values. (Here it can be taken that either $c_1 = c_2 = 1$ ppt or another scheme can be used for the computation of c_1, c_2.)

The integration was performed with $\Delta x = 1000$ m and $\Delta t = 100$ s, leading to a $V\Delta t/\Delta x = 0.1$.

A comparison of the results of the backward finite differences scheme (containing considerable numerical diffusion due to the small $V\Delta t/\Delta x$ value) with the analytical solution of Equation 6.12 is presented graphically in Fig. 6.1.

It is evident that this numerical scheme cannot describe advective–diffusive processes characterised by Pe > 10. An increase of the $V\Delta t/\Delta x$ value reduces the discrepancy between analytical and

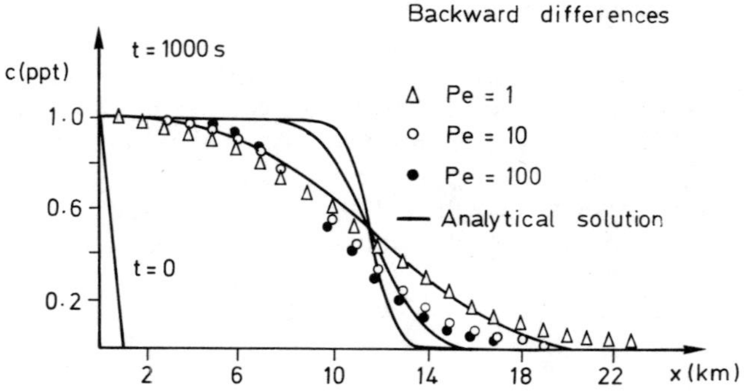

104 ADVECTIVE DIFFUSION AND DISPERSION

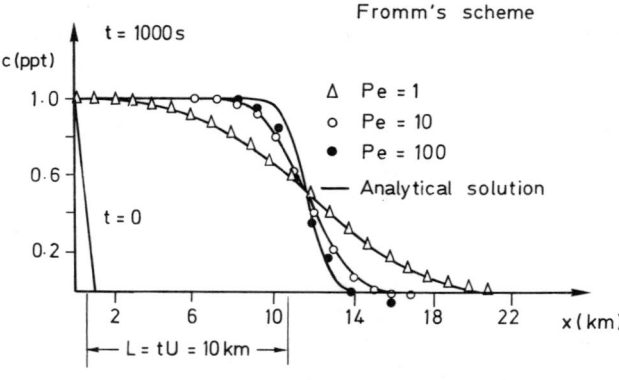

Fig. 6.1

Fig. 6.2

numerical results by reducing the numerical diffusion. A comparison between the results of Fromm's scheme and the analytical solution included in Fig. 6.2 shows that this scheme contains reduced numerical diffusion and can describe advective–diffusive processes up to $Pe = 100$.

7
The method of finite elements

7.1. HISTORICAL BACKGROUND—AN INTRODUCTION TO THE METHOD

The method of finite differences, presented earlier (Sections 1.6, 2.2–2.4), is applied where an analytical solution of the mathematical model does not exist, due to the difficulty of the differential equation to be solved or irregularities in the shape of the solution region. In the method of finite differences the solution region is envisaged as an ensemble of grid points and the differential equations are replaced by difference equations with respect to the function values at the grid points. The problem is thus transformed to an algebraic one.

There is an alternative approach, to envisage the solution region as an ensemble of interconnected subregions (finite elements), to solve the equation inside each element (following either the variational approach or that of weighted residuals, as will be seen later) and proceed with the assembly of the elements, which finally leads again to an algebraic problem.

The FE methodology and the developed numerical technique is a byproduct of research in statics and elasticity where there is a tendency to break down a complex structure to simpler elementary discs. The first papers appeared in 1954. An application of FE in plane elasticity appeared in 1960 and from that time on the method has been widely used in problems of continuum mechanics, aeroelasticity, applied mathematics, etc. Extended historical notes on the subject are included in Chung[6], Huebner[13] and Zienkiewicz[27]. The method can be used equally well in equilibrium, eigenvalue, and propagation problems.

The steps followed in the application of the method can be summarised as follows:

(1) Discretisation of the continuum into elements. The selection of the shape and size of the elements is a skill which can only be acquired through practice. Of course basic principles exist, as for example the use of smaller elements in areas where the field variable gradient increases sharply, but only experience can lead to the

optimal discretisation which in turn can radically influence the precision of the results.

(2) Selection of interpolation (shape) functions. The field variable is expressed inside each element in terms of the approximating (shape or base) functions. These functions are multiplied by the values of field variables on the nodal points of each element.

(3) Finding the element properties or, alternatively, solving the field equation for each element using variational calculus or the method of weighted residuals (which will be discussed later).

(4) Assembling the elements and forming and solving the system of algebraic equations derived.

Two basic questions are usually posed by a newcomer to the subject; Why finite elements and what is the relation between finite elements and finite differences? The answers vary according to personal experience which determines the degree of enthusiasm in supporting the claims of one method as opposed to the other. A moderate user of both FD and FE methods would give the following answers: finite elements originated and are used in solution domains of a complex geometric form, where difficulties occur in approximating irregular boundaries by finite difference grids. Both the FE and FD techniques replace the differential operators by algebraic equations having as unknowns the field variable values at characteristic locations.

Recent publications have shown a close relation between the FD and FE methods (Gray and Pinder[28]). FD and FE equations of the same form can be derived by use of weighted differences while the FE equations can be derived without the bias of weighting coefficients, based on error minimisation principles. One or other method is preferred according to the specific situation and no dogmatic conclusion favouring either method can be reached.

7.2. THE RITZ AND GALERKIN METHODS FOR THE APPROXIMATE SOLUTION OF DIFFERENTIAL EQUATIONS

7.2.1. Some principles of the calculus of variations

The theory of variations forms a self contained and interesting part of the classical calculus and it is not possible to present it here in detail. The essence of the theory can be summarised as follows: Variational calculus is used to investigate alternative formulations of field problems (continuum mechanics, electromagnetic theory, etc.) by means of stationary functionals. It can be proved that there is a one to one correspondence and equivalence between some differential equations and functionals which become stationary (a relative maximum or minimum value) for those functions only, which are solutions of the original differential equations. A functional is defined as the integral over the solution region of an unknown function and its derivatives,

$$I\{\varphi\} = \int_D F(x, \varphi, \varphi' \ldots) \, dx \qquad (7.1)$$

The goal is to find the function φ that makes the functional I stationary. It can be proved that the function φ that makes $I\{\varphi\}$ stationary is the solution of a differential equation called the Euler equation of the functional. The form of the Euler equations of some common functionals is given below:

$$I\{\varphi\} = \int_D F(x, \varphi, \varphi') \, dx \leftrightarrow F_\phi - \frac{d}{dx} F_{\phi'} = 0 \qquad (7.2)$$

where the subscripts denote partial derivative ($F_\phi = \partial F / \partial \varphi$)

$$I\{\varphi\} = \int_D F(x, \varphi, \varphi', \varphi'') \, dx \leftrightarrow F_\phi - \frac{d}{dx} F_{\phi'} + \frac{d^2}{dx^2} F_{\phi''} = 0 \qquad (7.3)$$

$$I\{\varphi\} = \iint_D F(x, y, \varphi, \varphi_x, \varphi_y) \, dx \, dy \leftrightarrow F_\phi - \frac{\partial}{\partial x} F_{\phi_x} - \frac{\partial}{\partial y} F_{\phi_y} = 0 \qquad (7.4)$$

As a simple example consider the functional

$$I\{\varphi\} = \int_l \left[\frac{1}{2} \left(\frac{d\varphi}{dx} \right)^2 + f\varphi \right] dx, \left(F = \frac{1}{2} \left(\frac{d\varphi}{dx} \right)^2 + f\varphi \right) \qquad (7.5)$$

The function that extremises it, must satisfy the Euler equation:

$$\frac{d^2\varphi}{dx^2} = f(x) \qquad (7.6)$$

which in one dimension is a simple linear second order ordinary differential equation and in two dimensions the Poisson equation.

The inverse process of finding a functional from the differential equation is not easy and not always possible. In many cases, however, there are definite advantages in handling the functional instead of the corresponding Euler equation, for the following reasons:

(1) The functional has a physical meaning, it describes, for example, the potential energy of the considered continuum.
(2) It contains lower order derivatives than the Euler equation, as shown by comparison of Equations 7.5 and 7.6.
(3) It takes care of some complicated boundary conditions in a simpler way.

A classical method for the approximate analytical solution of functional extremum problems will be given in detail as it is considered the precursor to the method of finite elements.

7.2.2. The Ritz method

The Ritz method is a classical method to obtain approximate solutions for problems expressed in variational form. According to this method the unknown function (the one that extremises the functional) is expressed in terms of a series of 'coordinate' functions multiplied by undetermined coefficients. The approximate form of the function is substituted in the functional, the integrations are performed and the algebraic expression with respect to the undetermined coefficients is extremised, that is, the partial derivatives of the expression with respect to the coefficients are made equal to zero. Thus, the extremum conditions form the final algebraic system with respect to the undetermined coefficients. Its solution gives the terms of the series approximating the unknown function. It should be noted that the coordinate functions are defined over the whole solution domain and must satisfy the boundary conditions of the problem.

As a simple illustration of the Ritz method let us consider the linear second order ordinary differential equation,

$$\frac{d^2\varphi}{dx^2} = f(x) = -x \tag{7.7}$$

with boundary conditions $\varphi|_{x=0} = 0$, $\varphi|_{x=1} = 0$. The functional having as Euler equation 7.7 is

$$I\{\varphi\} = \int_0^1 \left[\frac{1}{2}\left(\frac{d\varphi}{dx}\right)^2 - x\varphi\right] dx \quad (7.8)$$

Assume that $\varphi(x)$ has the general form $\hat{\varphi} = x(x-1)(c_1 + c_2 x + c_3 x^2 + \cdots)$. As a first step assume that only one term of the series is retained,

$$\hat{\varphi}^{(1)} = x(x-1)c_1 \quad (7.9)$$

Substitution of $\hat{\varphi}^{(1)}$ in I leads to the expression,

$$I\{\hat{\varphi}\} = \int_0^1 2c_1^2 x^2 + \frac{c_1^2}{2} - 2c_1^2 x - c_1 x^3 + c_1 x^2 = \text{stationary} \quad (7.10)$$

which after integration becomes,

$$I(c_1) = c_1^2(\tfrac{2}{3} + \tfrac{1}{2} - 1) + c_1(\tfrac{1}{3} - \tfrac{1}{4}) = \text{stationary} \quad (7.11)$$

For an extreme value the following condition must be satisfied

$$dI/dc_1 = 0 \quad (7.12)$$

giving

$$c_1 = -\tfrac{1}{4}$$

So, the first approximation of the investigated function satisfying Equation 7.7 and extremizing Equation 7.8 is,

$$\hat{\varphi}^{(1)} = x(1-x)/4 \quad (7.13)$$

Let us approximate φ to a greater degree of accuracy by $\hat{\varphi}^{(2)} = x(x-1)(c_1 + c_2 x)$. The functional now takes the form

$$I(c_1, c_2) = \int_0^1 \left[\tfrac{1}{2}(2c_1 x - c_1 + 3c_2 x^2 - 2c_2 x)^2 - \right.$$

$$\left. - x(c_1 x^2 - c_1 x + c_2 x^3 - c_2 x^2)\right] dx \quad (7.14)$$

and after integration

$$I(c_1, c_2) = \frac{c_1^2}{6} + \frac{c_2^2}{15} + \frac{c_1 c_2}{6} + \frac{c_1}{12} + \frac{c_2}{20} = \text{stationary} \quad (7.17)$$

A stationary value is achieved by means of the conditions

$$\partial I/\partial c_1 = \frac{2c_1}{6} + \frac{c_2}{6} + \frac{1}{12} = 0 \quad (7.16)$$

$$\partial I/\partial c_2 = \frac{2c_2}{15} + \frac{c_1}{6} + \frac{1}{20} = 0 \quad (7.17)$$

giving $c_1 = -\frac{1}{6}$, $c_2 = -\frac{1}{6}$ and a $\hat{\varphi}^{(2)}$ form,

$$\hat{\varphi}^{(2)} = x(x+1)(1-x)/6 \quad (7.18)$$

The analytical solution of Equation 7.7 with the given boundary conditions is

$$\varphi = x(1-x)(1+x)/6 \quad (7.19)$$

coinciding with the second order approximate solution given by the Ritz method.

7.2.3. The method of weighted residuals

This is the name given to a variety of approximate methods for the solution of partial differential equations. All of the methods are characterised by a common process described briefly as follows.

The unknown function is expressed in terms of a series of base functions multiplied by unknown coefficients. The series is substituted in the differential equation (not necessarily a functional). The differential equation is not satisfied since a residual (Re) appears, and the unknown coefficients are computed through a residual minimisation process. The method under many different forms and names is lucidly presented in a review paper by Finlayson and Scriven[29].

A brief mathematical illustration is given. Assume a differential operator $\mathscr{L}$ generating the equation

$$\mathscr{L}(\varphi) = f \quad (7.20)$$

where $\varphi = \varphi(x, y, z)$ is defined in a domain D bounded by a surface Γ. The equation is completed by appropriate boundary conditions on Γ.
The method of weighted residuals is analysed in the following steps:

(1) Assume a functional form of the field variable

$$\hat{\varphi} = \sum_{i=1}^{n} N_i c_i \qquad (7.21)$$

where $N_i(x, y, z)$ are base functions satisfying (at least approximately) the differential equation and the boundary conditions and c_i are undetermined coefficients which can be either constants or functions of time (in the case of time dependent problems).

(2) As the approximate form of $\hat{\varphi}$ does not satisfy the differential equation the substitution of $\hat{\varphi}$ in Equation 7.20 generates a residual

$$\mathrm{Re} = \mathscr{L}(\hat{\varphi}) - f \qquad (7.22)$$

The optimisation in the approximation of $\hat{\varphi}$ to the exact solution φ is achieved through the minimisation of the residual. According to the method of weighted residuals minimisation requires the 'orthogonality' of Re to a set of weighting functions,

$$\int_D (\mathscr{L}(\hat{\varphi}) - f) W_i \, \mathrm{d}D = 0, \quad i = 1 \ldots n \qquad (7.23)$$

(3) Integration leads to n algebriac equations, usually coupled to a system of simultaneous equations with respect to the n unknown coefficients C_i. The solution of the system gives the $\hat{\varphi}$ function in open form.

It can be proved mathematically that $\hat{\varphi}$ converges to φ with proper selection of N_i, W_i and increasing n. It is well known that if W_i are functions of a complete set and $n \to \infty$, then the only function orthogonal to all the elements of a complete set is zero.

Special names are assigned to the method according to the form of the W_i functions. For $W_i = N_i$, the method is known as the Galerkin method. A numerical example will be given in detail as the Galerkin version of the method of weighted residuals is widely applied to the FE method.

Assume the same differential equation, Equation 7.7, as in the previous example. According to the weighted residuals method the

112 THE METHOD OF FINITE ELEMENTS

function φ can be approximated by the following trigonometric series:

$$\hat{\varphi} = \sum_{k=1}^{n} c_k \sin k\pi x \tag{7.24}$$

The base function $\sin k\pi x$ satisfies the boundary conditions $\varphi|_{x=0} = 0$, $\varphi|_{x=1} = 0$. Let us solve the problem to a first approximation.

Substitution of Equation 7.24 in Equation 7.7 generates the residual

$$\text{Re} = -c_1 \pi^2 \sin \pi x + x \tag{7.25}$$

The minimisation equation according to the Galerkin method is,

$$\int_0^1 \text{Re} \sin \pi x \, dx = \int_0^1 (c_1 \pi^2 \sin \pi x - x) \sin \pi x \, dx = 0 \tag{7.26}$$

which after integration is solved for c_1, giving $c_1 = 2/\pi^3$.

As a second approximation

$$\hat{\varphi}^{(2)} = c_1 \sin \pi x + c_2 \sin 2\pi x \tag{7.27}$$

The residual has the form

$$\text{Re} = -c_1 \pi^2 \sin \pi x - 4c_2 \pi^2 \sin 2\pi x + x \tag{7.28}$$

and the Galerkin principle generates the two minimisation equations

$$\int_0^1 \text{Re} \sin \pi x \, dx = 0 \tag{7.29}$$

$$\int_0^1 \text{Re} \sin 2\pi x \, dx = 0 \tag{7.30}$$

which give $c_1 = 2/\pi^3$, $c_2 = -1/4\pi^3$.

Comparison with the analytical solution is depicted in Fig. 7.1.

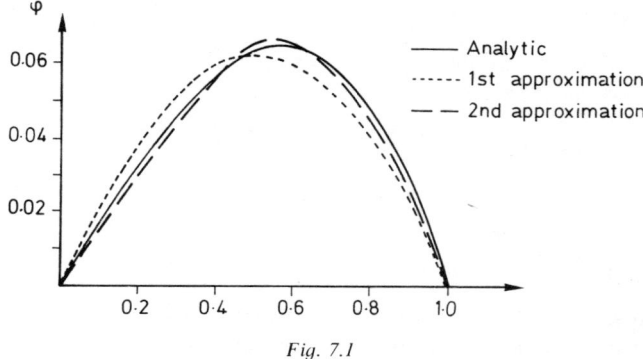

Fig. 7.1

7.3. DISCRETISATION BY FINITE ELEMENT SHAPE FUNCTIONS

In both the Ritz and Galerkin methods the base or coordinate functions are defined and extended over the solution domain and they all contribute to the computation of the values of the unknown function (field variable) at any point. The idea of using base functions defined over only a part of the solution domain is the opening step in the method of finite elements.

First, let us consider the domain discretisation. From the FD method the reader is accustomed to the establishment of a grid over the solution domain. The meshes are line segments, squares, parallelograms, cubes or orthogonal parallelepipeds for 1-D, 2-D and 3-D problems respectively.

A more relaxed approach for the discretisation of a 2-D field, for example, would be to use unequal triangles with rectilinear or curved sides, or any other geometric form. As soon as the coordinates of the vertices of a triangular element are known, the element is well defined.

The discretisation of the solution domain into elements of arbitrary shape and dimensions is a unique property of the method of finite elements. The spatial variation of the size of the elements is a simple graphic process and the passage from a coarse to a finer grid does not introduce any computational problems, as in the FD method.

Linear segments (1-D problems) and rectilinear triangular finite elements (2-D problems) will be described in more detail here, although many other elements have been proposed and described, e.g. Chung[6], Heubner[13] and Zienkiewicz[27].

A second characteristic of the FE method is that the base functions used for the approximation of the field variable are defined locally,

114 THE METHOD OF FINITE ELEMENTS

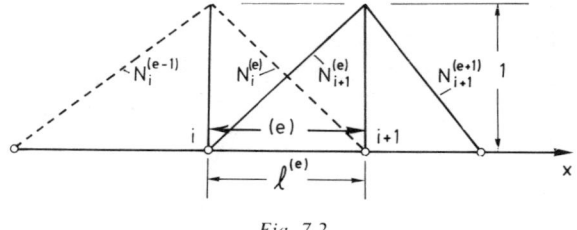

Fig. 7.2

extending over some elements only. Several functions may be used, the linear function being the most simple. The variation of the field variable over an element is expressed in terms of the local base or shape functions and the field variable values at characteristic locations on the element, a process known as piecewise approximation. Simple examples for 1-D and 2-D cases will illustrate these ideas.

Assume a 1-D problem posed on a solution domain extending along the x axis. The domain is divided into linear elements. A typical element (e) is defined by the nodal points i and $i+1$ at a distance $l^{(e)}$ (the length of the element). The shape functions to be used are linear functions defined on two neighbouring elements. The shape functions N_i, N_{i+1} have the form depicted in Fig. 7.2 and take a value equal to 1 at the nodal points $i, i+1$.

Their analytical form along the length of the element (e) are

$$N_i^{(e)}(x) = \frac{x_{i+1} - x_i}{l_i^{(e)}}, \quad \frac{\partial N_i^{(e)}}{\partial x} = -\frac{1}{l_i^{(e)}} \tag{7.31}$$

$$N_{i+1}^{(e)}(x) = \frac{x - x_i}{l_i^{(e)}}, \quad \frac{\partial N_{i+1}^{(e)}}{\partial x} = \frac{1}{l_i^{(e)}} \tag{7.32}$$

The field variable along the element (e) can be expressed by means of the shape functions N_i, N_{i+1} and the nodal values of field variable φ_i, φ_{i+1} as:

$$\hat{\varphi}^{(e)} = N_i^{(e)} \varphi_i + N_{i+1}^{(e)} \varphi_{i+1} = [N_i^{(e)}, N_{i+1}^{(e)}] \begin{Bmatrix} \varphi_i \\ \varphi_{i+1} \end{Bmatrix} \tag{7.33}$$

or in matrix form

$$\hat{\varphi}^{(e)} = [N]^{(e)} \{\varphi\}^{(e)} \tag{7.34}$$

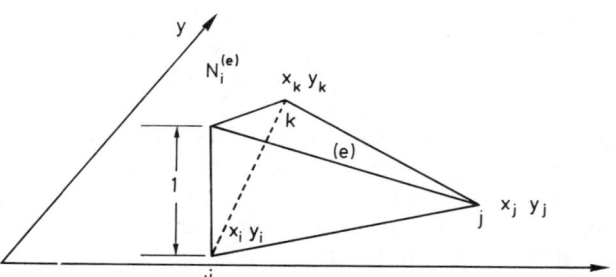

Fig. 7.3

Along the whole solution domain, divided into M elements, the function φ is expressed approximately by the sum:

$$\hat{\varphi} = \sum_{e=1}^{M} \varphi^{(e)} = \sum_{e=1}^{M} [N]^{(e)}\{\varphi\}^{(e)} \tag{7.35}$$

It is evident that a local value $\varphi(x)$, where $x_i \leqslant x \leqslant x_{i+1}$, is influenced only by values of the base functions and nodal φ values extending from x_i to x_{i+1}. This is the essence of piecewise approximation.

In the case of 2-D problems the same approach is followed. Here, a simpler form is the rectilinear triangular element (e) defined by coordinates of the vertices $(x_i y_i, x_j y_j, x_k y_k)$.

A piecewise approximation of the field variable φ can be achieved through the nodal values and linear base functions $N_i^{(e)}, N_j^{(e)}, N_k^{(e)}$ having the form of pyramids (hat or chapeau functions) depicted in Fig. 7.3 and expressed analytically by the equations:

$$N_i^{(e)} = \frac{a_i^{(e)} + b_i^{(e)} x + c_i^{(e)} y}{2\Delta} \tag{7.36}$$

where

$$a_i^{(e)} = x_j y_k - x_k y_j; \quad b_i^{(e)} = y_j - y_k; \quad c_i^{(e)} = x_k - x_j \tag{7.37}$$

and

$$2\Delta = \begin{vmatrix} 1 & x_i & y_i \\ 1 & x_j & y_j \\ 1 & x_k & y_k \end{vmatrix} \tag{7.38}$$

Similarly, the N_j, N_k base functons can be expressed through cyclic permutation of the indices $i \to j \to k \to i$.

The $\hat{\varphi}^{(e)}$ expression is again:

$$\hat{\varphi}^{(e)} = N_i^{(e)}\varphi_i + N_j^{(e)}\varphi_j + N_k^{(e)}\varphi_k = [N]^{(e)}\{\varphi\}^{(e)} \qquad (7.39)$$

For higher order shape functions, and non-linear variation of the field variable over the element, more than three nodal points are required, usually spread over the sides of the triangular elements. Elements of other than triangular form and higher order shape functions are described in Chung[6].

7.4. DERIVATION OF FINITE ELEMENT EQUATIONS BY THE RITZ OR GALERKIN METHOD

The application of the approximate forms $\hat{\varphi}^{(e)}$ to the differential equation or the equivalent functional and the use of the Galerkin or Ritz methods lead to equations for the element (e) containing as unknowns the nodal values φ_i, φ_{i+1} or φ_i, φ_j, φ_k for one- and two-dimensional cases respectively. The Galerkin method, for example, can be formally written for the shape function N_i extending along elements (e) and ($e-1$) of the 1-D problem,

$$\int_{(e),(e-1)} [\mathscr{L}(\hat{\varphi}) - f] N_i \, dx = 0 \qquad (7.40)$$

where

$$\hat{\varphi} = [N]^{(e-1)}\{\varphi\}^{(e-1)} + [N]^{(e)}\{\varphi\}^{(e)}$$

$$= [N_{i-1}\, N_i]\begin{Bmatrix}\varphi_{i-1}\\ \varphi_i\end{Bmatrix} + [N_i\, N_{i+1}]\begin{Bmatrix}\varphi_i\\ \varphi_{i+1}\end{Bmatrix} \qquad (7.41)$$

From the above summation it is concluded that the formulation of the FE equation is performed as follows: The elemental contributions

$$\int_{(e)} [\mathscr{L}(\hat{\varphi})^{(e)} - f^{(e)}] N_k^{(e)} \, dx, \quad k = i, i+1 \qquad (7.42)$$

are formed and then, by proper addition, the global equations for all the weighting functions $W_i \equiv N_i$ are formed as in Equation 7.35. These equations have the form of a system of simultaneous algebraic equations for the nodal values φ_i.

The global system has a matrix with elements inserted automatically, as they are formulated. The assembly procedure can be programmed easily as will be shown in the applications. The matrices are usually sparse and special procedures have been devised for their inversion (solution of the system).

As can be seen from the form of Equation 7.42 the estimation of the integral of the shape functions and their products are needed for the computation of the elemental contribution in the global equations. A short list of formulae used in 1-D and 2-D (linear triangular elements) is given.

1-Dimensional elements (line segments)

For an element (e) defined by points $i, i+1$ on x axis

$$\hat{\varphi}^{(e)}(x) = N_i^{(e)}\varphi_i + N_{i+1}^{(e)}\varphi_{i+1} = [N]^{(e)}\{\varphi\}^{(e)} \tag{7.43}$$

where

$$N_i^{(e)} = \frac{x_{i+1} - x}{l^{(e)}}, \quad N_{i+1}^{(e)} = \frac{x - x_i}{l^{(e)}}, \quad \frac{\partial N_i^{(e)}}{\partial x} = -\frac{1}{l^{(e)}}, \quad \frac{\partial N_{i+1}^{(e)}}{\partial x} = \frac{1}{l^{(e)}} \tag{7.44}$$

and

$$\int_{l^{(e)}} N_i^\alpha N_{i+1}^\beta \, dx = \frac{\alpha!\beta!l^{(e)}}{(\alpha+\beta+1)!} \tag{7.45}$$

2-Dimensional elements (rectilinear triangles)

For an element (e) defined by the vertices i, j, k on the x, y plane

$$\hat{\varphi}^{(e)}(x, y) = N_i^{(e)}\varphi_i + N_j^{(e)}\varphi_j + N_k^{(e)}\varphi_k = [N]^{(e)}\{\varphi\}^{(e)} \tag{7.46}$$

where

$$N_i^{(e)} = \frac{1}{2\Delta}(a_i + b_i x + c_i y), \quad \frac{\partial N_i^{(e)}}{\partial x} = \frac{b_i}{2\Delta}, \quad \frac{\partial N_i^{(e)}}{\partial y} = \frac{c_i}{2\Delta} \tag{7.47}$$

118   THE METHOD OF FINITE ELEMENTS

$$N_j^{(e)} = \frac{1}{2\Delta}(a_j + b_j x + c_j y), \frac{\partial N_j^{(e)}}{\partial x} = \frac{b_j}{2\Delta}, \frac{\partial N_j^{(e)}}{\partial y} = \frac{c_j}{2\Delta} \quad (7.48)$$

$$N_k^{(e)} = \frac{1}{2\Delta}(a_k + b_k x + c_k y), \frac{\partial N_k^{(e)}}{\partial x} = \frac{b_k}{2\Delta}, \frac{\partial N_k^{(e)}}{\partial y} = \frac{c_k}{2\Delta} \quad (7.49)$$

with

$$\Delta = \tfrac{1}{2} \det \begin{bmatrix} 1 & x_i & y_i \\ 1 & x_j & y_j \\ 1 & x_k & y_k \end{bmatrix} \quad (7.50)$$

$$a_i = x_j y_k - x_k y_j, \quad b_i = y_j - y_k, \quad c_i = x_k - x_j \quad (7.51)$$

The remainder of the coefficients are obtained by cycling the indices, $i \rightarrow j \rightarrow k \rightarrow i$

$$\int_{(e)} N_i^\alpha N_j^\beta N_k^\gamma \, dx \, dy = \frac{\alpha! \beta! \gamma! 2\Delta}{(\alpha + \beta + \gamma + 2)!} \quad (7.52)$$

As an illustration let us consider the formulation and solution of FE equations by means of the Ritz and the Galerkin method for a typical ordinary differential equation.

7.4.1. Application of the Ritz procedure

For the previously considered 1-D problem of the second order linear ordinary differential equation, Equation 7.7, let us apply first the Ritz method with the solution domain extending from $x=0$ to 1, discretized into 2 and 3 elements.

For 2 equal elements, $l_1 = l_2 = 0.5$, and linear shape functions N_1, N_2, N_3, as shown in Fig. 7.4, the functional is written as

$$I\{\varphi\} = \int_0^1 \left[\frac{1}{2}\left(\frac{d\varphi}{dx}\right)^2 - \varphi x \right] dx =$$

$$= \int_0^{1/2} \frac{1}{2}\left(\frac{d}{dx}[N_1^{(1)} N_2^{(1)}] \begin{Bmatrix} \varphi_1 \\ \varphi_2 \end{Bmatrix} \right)^2 dx - \int_0^{1/2} [N_1^{(1)} N_2^{(1)}] \begin{Bmatrix} \varphi_1 \\ \varphi_2 \end{Bmatrix} x \, dx +$$

THE METHOD OF FINITE ELEMENTS

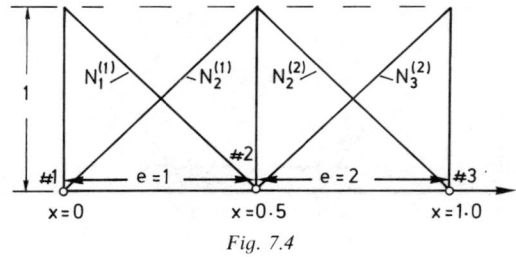

Fig. 7.4

$$+ \int_{1/2}^{1} \frac{1}{2}\left(\frac{d}{dx}[N_2^{(2)} N_3^{(2)}]\begin{Bmatrix}\varphi_2\\\varphi_3\end{Bmatrix}\right)^2 dx - \int_{1/2}^{1} [N_2^{(2)} N_3^{(2)}]\begin{Bmatrix}\varphi_2\\\varphi_3\end{Bmatrix} x\, dx =$$

$$= \text{stationary} \tag{7.53}$$

For

$$N_1^{(1)} = \frac{0.5 - x}{0.5}, \quad \frac{dN_1^{(1)}}{dx} = -2 \tag{7.54}$$

$$N_2^{(1)} = \frac{x}{0.5}, \quad \frac{dN_2^{(1)}}{dx} = 2, \quad N_2^{(2)} = \frac{1-x}{0.5}, \quad \frac{dN_2^{(2)}}{dx} = -2 \tag{7.55}$$

$$N_3^{(2)} = \frac{x - 0.5}{0.5}, \quad \frac{dN_3^{(2)}}{dx} = 2 \tag{7.56}$$

and $\varphi_1 = \varphi_3 = 0$, according to the given boundary conditions, the above functional becomes

$$I\{\varphi_2\} = \int_0^{1/2} \tfrac{1}{2}(2\varphi_2)^2\, dx - \int_0^{1/2} 2x^2 \varphi_2\, dx + \int_{1/2}^{1} \tfrac{1}{2}(-2\varphi_2)^2\, dx -$$

$$- \int_{1/2}^{1} 2(1-x) x \varphi_2\, dx \tag{7.57}$$

or

$$I\{\varphi_2\} = 2\varphi_2^2 - 0.083 \varphi_2 - 0.167 \varphi_2 \tag{7.58}$$

120 THE METHOD OF FINITE ELEMENTS

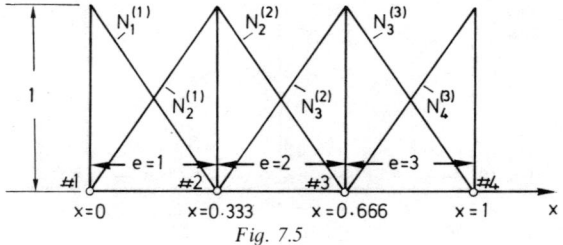

Fig. 7.5

A stationary value implies

$$\frac{dI}{d\varphi_2} = 0 \rightarrow \varphi_2 = 0.063 \tag{7.59}$$

For 3 elements $l_1 = l_2 = l_3 = 0.333$ and linear shape functions N_1 to N_4, given in Fig. 7.5, the functional takes the form:

$$I\{\varphi\} = \int_0^{1/3} \frac{1}{2}\left(\frac{d}{dx}[N_1 N_2]^{(1)}\begin{Bmatrix}\varphi_1\\\varphi_2\end{Bmatrix}\right)^2 dx - \int_0^{1/3} [N_1 N_2]^{(1)}\begin{Bmatrix}\varphi_1\\\varphi_2\end{Bmatrix} x \, dx +$$

$$+ \int_{1/3}^{2/3} \frac{1}{2}\left(\frac{d}{dx}[N_2 N_3]^{(2)}\begin{Bmatrix}\varphi_2\\\varphi_3\end{Bmatrix}\right)^2 dx - \int_{1/3}^{2/3} [N_2 N_3]^{(2)}\begin{Bmatrix}\varphi_2\\\varphi_3\end{Bmatrix} x \, dx +$$

$$+ \int_{2/3}^{1} \frac{1}{2}\left(\frac{d}{dx}[N_3 N_4]^{(3)}\begin{Bmatrix}\varphi_3\\\varphi_4\end{Bmatrix}\right)^2 dx - \int_{2/3}^{1} [N_3 N_4]^{(3)}\begin{Bmatrix}\varphi_3\\\varphi_4\end{Bmatrix} x \, dx \tag{7.60}$$

Substituting the shape functions N_1 to N_4 by the expressions

$$N_1^{(1)} = (0.333 - x)/0.333, \quad N_2^{(1)} = x/0.333 \tag{7.61}$$

$$N_2^{(2)} = (0.666 - x)/0.333, \quad N_3^{(2)} = (x - 0.333)/0.333 \tag{7.62}$$

$$N_3^{(3)} = (1 - x)/0.333, \quad N_4^{(3)} = (x - 0.666)/0.333 \tag{7.63}$$

and taking into account that $\varphi_1 = \varphi_4 = 0$, after integration the functional becomes a function of φ_2, φ_3,

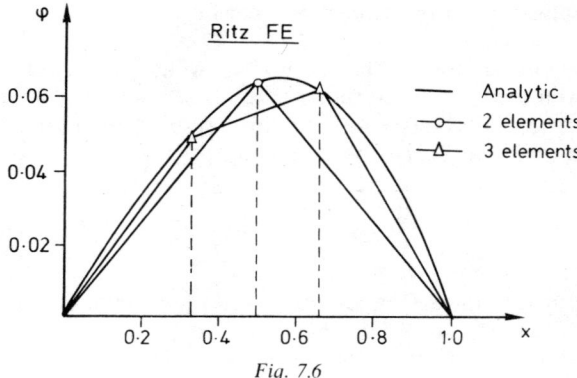

Fig. 7.6

$$I(\varphi_2, \varphi_3) = \int_0^{1/3} \tfrac{1}{2}(3\varphi_2)^2 \, dx + \int_{1/3}^{2/3} \tfrac{1}{2}(-3\varphi_2 + 3\varphi_3)^2 \, dx +$$

$$+ \int_{2/3}^{1} \tfrac{1}{2}(-3\varphi_3)^2 \, dx - \int_0^{1/3} 3\varphi_2 x^2 \, dx -$$

$$- \int_{1/3}^{2/3} 3[(\tfrac{2}{3}-x)\varphi_2 + (x-\tfrac{1}{3})\varphi_3] x \, dx - \int_{2/3}^{1} 3\varphi_3(1-x)x \, dx$$

$$= 3\varphi_2^2 + 3\varphi_3^2 - 3\varphi_2\varphi_3 - 0.112\varphi_2 - 0.221\varphi_3 \tag{7.64}$$

The stationarity is achieved through the conditions

$$\partial I/\partial \varphi_2 = 0 \rightarrow 6\varphi_2 - 3\varphi_3 = 0.112 \tag{7.65}$$

$$\partial I/\partial \varphi_3 = 0 \rightarrow -3\varphi_2 + 6\varphi_3 = 0.221 \tag{7.66}$$

The solution of the system gives $\varphi_2 = 0.049$, $\varphi_3 = 0.062$.

A graphic comparison of the analytical to the FE solutions for 2 and 3 elements is shown in Fig. 7.6.

7.4.2. Application of the Galerkin procedure

The Galerkin version of the method of weighted residuals can be equally well applied to the same situation. A small problem arises with the second derivative of the field variable which appears in the form of the second derivative of the shape function

$$\frac{d^2\hat{\varphi}}{dx^2} = \left(\frac{d^2}{dx^2}[N]\right)\{\varphi\} \qquad (7.67)$$

In the case of linear shape functions, their second derivative is identically equal to zero. This problem does not appear with the Ritz method as the functionals contain lower order derivatives than the equivalent differential equations. This difficulty can be overcome through integration by parts (or the application of Green's theorem for 2-D and 3-D problems). It is known that

$$\int_{x_1}^{x_2} \frac{dA}{dx} B \, dx = AB\Big|_{x_1}^{x_2} - \int_{x_1}^{x_2} A \frac{dB}{dx} \, dx \qquad (7.68)$$

If this procedure is applied to a finite element, apart from the change of form of the integrand, a new term containing boundary values of the field variable appears on the right hand side. These new terms from neighbouring elements are mutually eliminated (as they appear with different signs) when the elemental contributions are assembled in the global equations. Thus only the values on the boundary of the solution domain remain, and these are subject to the boundary conditions of the problem.

The Galerkin method (after integration by parts) is applied to the previous problem for 2 and 3 elements. For 2 elements the residual minimisation takes the form:

$$\int_0^1 \operatorname{Re} N_2 \, dx = -\int_0^{1/2} \frac{d}{dx}[N_1 N_2]^{(1)} \begin{Bmatrix} \varphi_1 \\ \varphi_2 \end{Bmatrix} \frac{d}{dx} N_2^{(1)} \, dx + \int_0^{1/2} x N_2^{(1)} \, dx -$$

$$- \int_{1/2}^1 \frac{d}{dx}[N_2 N_3]^{(2)} \begin{Bmatrix} \varphi_2 \\ \varphi_3 \end{Bmatrix} \frac{d}{dx} N_2^{(2)} \, dx +$$

$$+ \int_{1/2}^{1} xN_2^{(2)} \, dx + \frac{d}{dx}[N_1N_2]^{(1)}\begin{Bmatrix}\varphi_1\\\varphi_2\end{Bmatrix}N_2^{(1)}|_0^{1/2} +$$

$$+ \frac{d}{dx}[N_2N_3]^{(2)}\begin{Bmatrix}\varphi_2\\\varphi_3\end{Bmatrix}N_2^{(2)}|_{1/2}^{1} = 0 \tag{7.69}$$

For N_1, N_2, N_3 expressed by Equations 7.54 to 7.56 and $\varphi_1 = \varphi_3 = 0$, Equation 7.69 takes the form,

$$-\int_0^{1/2}(2\varphi_2)2 \, dx + \int_0^{1/2} x \cdot 2x \, dx -$$

$$-\int_{1/2}^{1}(-2\varphi_2)(-2) \, dx + \int_{1/2}^{1} x(1-x)2 \, dx = 0 \tag{7.70}$$

giving, after integration, $\varphi_2 = 1/16$.

For 3 equal elements the unknown nodal values are φ_2, φ_3 and two equations for the residual minimisation are formed.

$$\int_{e_1+e_2} \text{Re } N_2 \, dx = -\int_0^{1/3} \frac{d}{dx}[N_1N_2]^{(1)}\begin{Bmatrix}\varphi_1\\\varphi_2\end{Bmatrix}\frac{d}{dx}N_2^{(1)} \, dx + \int_0^{1/3} xN_2^{(1)} \, dx -$$

$$-\int_{1/3}^{2/3} \frac{d}{dx}[N_2N_3]^{(2)}\begin{Bmatrix}\varphi_2\\\varphi_3\end{Bmatrix}\frac{d}{dx}N_2^{(2)} \, dx + \int_{1/3}^{2/3} xN_2^{(2)} \, dx = 0 \tag{7.71}$$

and

$$\int_{e_2+e_3} \text{Re } N_3 \, dx = -\int_{1/3}^{2/3} \frac{d}{dx}[N_2N_3]^{(2)}\begin{Bmatrix}\varphi_2\\\varphi_3\end{Bmatrix}\frac{d}{dx}N_3^{(2)} \, dx + \int_{1/3}^{2/3} xN_3^{(2)} \, dx -$$

$$-\int_{2/3}^{1} \frac{d}{dx}[N_3N_4]^{(3)}\begin{Bmatrix}\varphi_3\\\varphi_4\end{Bmatrix}\frac{d}{dx}N_3^{(3)} \, dx + \int_{2/3}^{1} xN_3^{(3)} \, dx = 0 \tag{7.78}$$

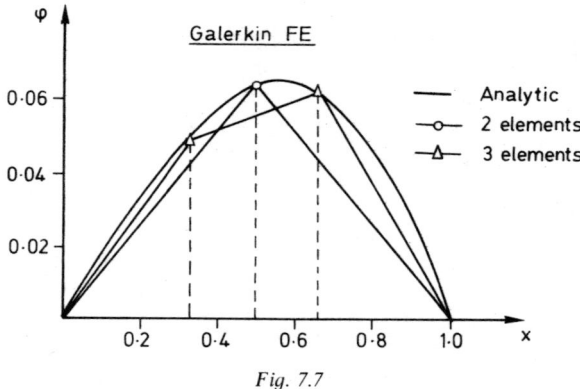

Fig. 7.7

After substitution of N_1, N_2, N_3, N_4 in Equations 7.61–7.63 and as $\varphi_1 = \varphi_4 = 0$, performing the integrations leads to the following coupled equations for φ_2, φ_3:

$$\varphi_3 - 2\varphi_2 = -0.01373 \tag{7.73}$$

$$\varphi_2 - 2\varphi_3 = -0.0740 \tag{7.74}$$

which gives $\varphi_2 = 0.049$ and $\varphi_3 = 0.0617$.

Comparison of the Galerkin FE solution to the analytical one is shown in Fig. 7.7.

For a larger number of elements a computer program can be formed to calculate the elemental contributions to the global equations and to solve the simultaneous algebraic equations. For the same problem, solved by the Galerkin FE method, a computer program can be composed on the basis that for an element (e) of length l defined by points x_i, x_{i+1} the elemental contributions to the global system

$$[A]\{\varphi\} = \{B\} \tag{7.75}$$

are:

$$A_{i,i} = \int_{(e)} \frac{d}{dx} N_i \frac{d}{dx} N_i \, dx = \frac{1}{l} \tag{7.76}$$

$$A_{i,i+1} = \int_{(e)} \frac{d}{dx} N_{i+1} \frac{d}{dx} N_i \, dx = -\frac{1}{l} \tag{7.77}$$

THE METHOD OF FINITE ELEMENTS

$$A_{i+1,i} = \int_{(e)} \frac{d}{dx} N_i \frac{d}{dx} N_{i+1} \, dx = -\frac{1}{l} \tag{7.78}$$

$$A_{i+1,i+1} = \int_{(e)} \frac{d}{dx} N_{i+1} \frac{d}{dx} N_{i+1} \, dx = \frac{1}{l} \tag{7.79}$$

$$B_i = \int_{(e)} x N_i \, dx = \int_{(e)} x \frac{x_{i+1} - x}{l} \, dx = \frac{x_{i+1}^2 - x_i^2}{2l} x_{i+1} - \frac{x_{i+1}^3 - x_i^3}{3l} \tag{7.80}$$

$$B_{i+1} = \int_{(e)} x N_{i+1} \, dx = \int_{(e)} x \frac{x - x_i}{l} \, dx = \frac{x_{i+1}^3 - x_i^3}{3l} - \frac{x_{i+1}^2 - x_i^2}{2l} x_i \tag{7.81}$$

For ten equal elements, $DX = 0.1$, $x_i = DX(i-1)$, the FORTRAN listing could have the form:

```
C     FINITE ELEMENT SOLUTION OF LINEAR 2ND ORDER ODE
      DIMENSION A(11,11),B(11),F(11)
      READ(5,1)DX,IMAX
    1 FORMAT(F7.0,I4)
      IMAX1=IMAX-1
      DO 2 I=1,IMAX
      F(I)=0.
      B(I)=0.
      DO 2 J=1,IMAX
    2 A(I,J)=0.
      DO 3 I=1,IMAX1
      A(I,I)=A(I,I)+1./DX
      A(I,I+1)=A(I,I+1)-1./DX
      A(I+1,I)=A(I+1,I)-1./DX
      A(I+1,I+1)=A(I+1,I+1)+1./DX
      B(I)=B(I)+DX**2*(I*(I**2-(I-1)**2)/2.-(I**3-
    1 (I-1)**3)/3)
      B(I+1)=B(I+1)+DX**2*((I**3-(I-1)**3)/3.-
    1 (I-1)*(I**2-(I-1)**2)/2.
    3 CONTINUE
      ITER=0
  100 ITER=ITER+1
      DIFMX=0.
      IF(ITER.GT.200) STOP 1
      DO 4 I=2,IMAX1
      TEMP=F(I)
      F(I)=(B(I)-A(I,I+1)*F(I+1)-A(I,I-1)*F(I-1))/
    1 A(I,I)
```

126 THE METHOD OF FINITE ELEMENTS

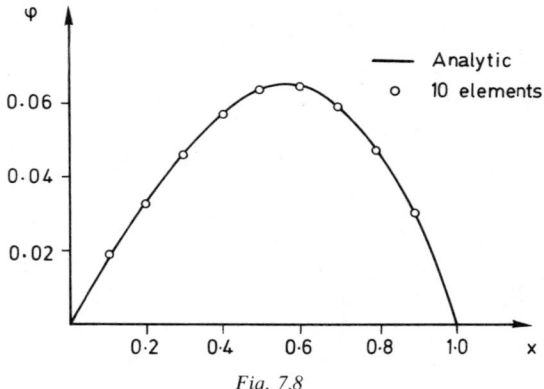

Fig. 7.8

```
      DIF=ABS(TEMP−F(I))
      IF(DIF.GT.DIFMX) DIFMX=DIF
 4    CONTINUE
      IF(DIFMX.GT..0001) GO TO 100
      WRITE(6,5)(F(I),I=1,IMAX)
 5    FORMAT(11F10.3)
      STOP
      END
```

For $DX = 0.1$ and $IMAX = 11$ the FE solution is graphically compared to the analytical one in Fig. 7.8.

Concluding the introduction to the FE method, the five steps which form the basis of the procedure are listed:
(1) Discretisation of the solution domain into elements.
(2) Selection of interpolation functions (shape functions).
(3) Formulation of elemental equation by means of variational weighted residuals methods.
(4) Assembly of the elemental equations into a global system.
(5) Solution of the algebraic system.

7.5. APPLICATION TO LONG LINEAR WAVES IN OPEN CHANNELS

The St. Venant equations for 1-D non-steady flow with respect to the free surface elevation ζ and the depth mean velocity u have the form:

$$\frac{\partial u}{\partial t} + u \frac{\partial u}{\partial x} = -g \frac{\partial \zeta}{\partial x} - \frac{\tau_b}{\rho h} \qquad (7.82)$$

$$\frac{\partial \zeta}{\partial t} + \frac{\partial}{\partial x}(hu) = 0 \tag{7.83}$$

where the bottom shear stress τ_b can be expressed as a function of the velocity. The usual expression is

$$\frac{\tau_b}{\rho} = ku|u| \tag{7.84}$$

but a linearised version

$$\frac{\tau_b}{\rho} = k'u \tag{7.85}$$

where k', an equivalent friction coefficient, is permissible.

For gradual variations in $\zeta(x, t)$ (propagation of long waves) and small variations in $h(x)$ the non-linear term $u\, \partial u/\partial x$ is usually neglected and the linearised version of Equations 7.82 and 7.83 is,

$$\frac{\partial u}{\partial t} = -g\frac{\partial \zeta}{\partial x} - \frac{k'}{h} u \tag{7.86}$$

$$\frac{\partial \zeta}{\partial t} + \frac{\partial}{\partial x}(hu) = 0 \tag{7.83}$$

The two equations, 7.86 and 7.83, can be combined to eliminate the u variable. The deriving equation with respect to ζ is the well known telegraphy equation (linear 2nd order hyperbolic equation)

$$\frac{\partial^2 \zeta}{\partial t^2} = g\frac{\partial}{\partial x}\left(h \frac{\partial \zeta}{\partial x}\right) - \frac{k'}{h}\frac{\partial \zeta}{\partial t} \tag{7.87}$$

This equation will be solved by the method of FE in a 1-D flow domain with constant bottom slope.

The initially horizontal free surface is perturbed by the passage of a sinusoidal wave generated at the left boundary while the right boundary is taken as a reflecting surface. The flow domain and the boundary conditions are depicted in Fig. 7.9.

The flow domain is discretised into equal elements of length DX, and linear shape functions are used. The water depth is assumed constant along each element (e), $h^{(e)} = $ const. The Galerkin finite element formulation of the problem (after integration by parts of the term $(\partial/\partial x)(h(\partial \zeta/\partial x))$ containing the second derivative of ζ is:

128 THE METHOD OF FINITE ELEMENTS

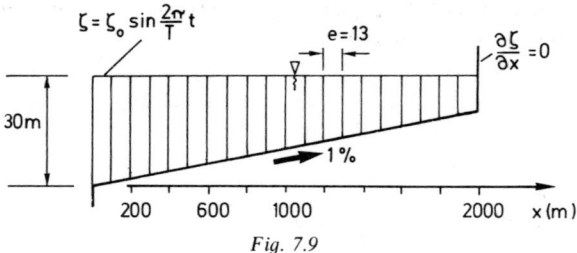

Fig. 7.9

$$\sum_{e=1}^{M} \left(\int_{(e)} [N]^{(e)} \left\{ \frac{\partial^2 \zeta}{\partial t^2} \right\} N_k^{(e)} \, dx + gh^{(e)} \int_{(e)} \frac{\partial}{\partial x} [N]^{(e)} \{\zeta\} \frac{\partial}{\partial x} N_k^{(e)} \, dx \right.$$

$$\left. + \frac{k'}{h^{(e)}} \int_{(e)} [N]^{(e)} \left\{ \frac{\partial \zeta}{\partial t} \right\} N_k^{(e)} \, dx \right) = 0 \quad (7.88)$$

where for the element (e), defined by the nodal points i and $i+1$,

$$[N] = [N_i, N_{i+1}], \quad \{\zeta\} = \left\{ \begin{array}{c} \zeta_i \\ \zeta_{i+1} \end{array} \right\}$$

and $k = i, i+1$.

The time derivatives appearing in Equation 7.88 are expressed by finite differences. If n is the time index ($t_n = n\Delta t$) the values of the field variables involved in the equation are:

$$\frac{\partial^2 \zeta}{\partial t^2} = (\zeta_i^{n+1} - 2\zeta_i^n + \zeta_i^{n-1})/\Delta t^2 \quad (7.89)$$

$$\frac{\partial \zeta}{\partial t} = (\zeta_i^n - \zeta_i^{n-1})/\Delta t \quad (7.90)$$

$$\zeta = \zeta_i^n \quad \text{(in the space derivatives)} \quad (7.91)$$

The scheme is explicit in time. Thus the algebraic system to be solved for each time step has the form

$$[A]\{\zeta^{n+1}\} = \{B(\zeta^n, \zeta^{n-1})\} \quad (7.92)$$

and the contributions of the element (e) to the left and right hand 2-D and 1-D matrices are respectively:

$$A_{ii} = \Delta x/3 \tag{7.93}$$

$$A_{i,i+1} = \Delta x/6 \tag{7.94}$$

$$A_{i+1,i} = \Delta x/6 \tag{7.95}$$

$$A_{i+1,i+1} = \Delta x/3 \tag{7.96}$$

$$\begin{aligned}
B_i &= (2\zeta_i^n - \zeta_i^{n-1})\Delta x/3 + (2\zeta_{i+1}^n - \zeta_{i+1}^{n-1})\Delta x/6 - \\
&\quad - gh^{(e)}\Delta t^2(\zeta_i^n - \zeta_{i+1}^n)/\Delta x - \\
&\quad - k'\Delta t^2/h^{(e)}[(\zeta_i^n - \zeta_i^{n-1})/3 + (\zeta_{i+1}^n - \zeta_{i+1}^{n-1})/6]\Delta x/\Delta t
\end{aligned} \tag{7.97}$$

$$\begin{aligned}
B_{i+1} &= (2\zeta_i^n - \zeta_i^{n-1})\Delta x/6 + (2\zeta_{i+1}^n - \zeta_{i+1}^{n-1})\Delta x/3 - \\
&\quad - gh^{(e)}\Delta t^2(-\zeta_i^n + \zeta_{i+1}^n)/\Delta x - \\
&\quad - k'\Delta t^2/h^{(e)}[(\zeta_i^n - \zeta_i^{n-1})/6 + (\zeta_{i+1}^n - \zeta_i^{n-1})/3]\Delta x/\Delta t
\end{aligned} \tag{7.98}$$

The ζ value on the left hand boundary is described by the equation:

$$\zeta_i^n = \zeta_0 \sin(2\pi t n/T) \tag{7.99}$$

and on the right hand boundary ($i = i_{\max}$) the reflection condition is

$$\zeta_{i_{\max}}^n = \zeta_{i_{\max}-1}^n \tag{7.100}$$

The FORTRAN program for the above FE problem is given below:

```
C     FINITE ELEMENT SOLUTION FOR LONG
C     LINEAR WAVES IN OPEN CHANNEL
      DIMENSION A(21,21),B(21),ZN(21),Z(21),ZO(21),H(21)
      READ(5,2)DX,DT,FRICT,PER,AMPL,IMAX,NMAX
    2 FORMAT(5F7.0,2I4)
      IMAX1=IMAX-1
      READ(5,3)(H(I),I=1,IMAX1)
    3 FORMAT(10F7.0)
      DO 4 I=1,IMAX
      ZN(I)=0.
      Z(I)=0.
```

```
    4   ZO(I)=0.
        T=0.
        N=0
  100   T=T+DT
        N=N+1
        ZN(1)=AMPL·SIN(T/PER·2832)
        DO 5 I=1,IMAX
        B(I)=0.
        DO 5 J=1,IMAX
    5   A(I,J)=0.
        DO 6 I=1,IMAX1
        B(I)=B(I)+((2*Z(I)−ZO(I))/3+(2*Z(I+1)−ZO(I+1))/6)
      1 *DX−9.81*H(I)*DT**2*(Z(I)−Z(I+1))/DX−
      1 FRICT*DT/H(I)*((Z(I)−ZO(I))/3+(Z(I+1)−
      1 ZO(I+1))/6)*DX
        B(I+1)=B(I+1)+((2*Z(I)−ZO(I))/6+(2*Z(I+1)−
      1 ZO(I+1))/3)
      1 *DX−9.81*H(I)*DT**2*(−Z(I)+Z(I+1))/DX−
      1 FRICT*DT/H(I)*((Z(I)−ZO(I))/6+(Z(I+1)−
      1 ZO(I+1))/3)*DX
        A(I,I)=A(I,I)+DX/3
        A(I,I+1)=A(I,I+1)+DX/6
        A(I+1,I)=A(I+1,I)+DX/6
    6   A(I+1,I+1)=A(I+1,I+1)+DX/3
        ITER=0
   99   ITER=ITER+1
        IF(ITER.GT.100) STOP 1
        DIFMX=0.
        DO 7 I=2,IMAX1
        TEMP=ZN(I)
        ZN(I)=(B(I)−ZN(I+1)*A(I,I+1)−ZN(I−1)
      1 *A(I,I−1))/A(I,I)
        DIF=ABS(TEMP−ZN(I))
        IF(DIF.GT.DIFMX)DIFMX=DIF
    7   CONTINUE
        IF(DIFMX.GT..0001) GO TO 99
        ZN(IMAX)=ZN(IMAX1)
        DO 8 I=1,IMAX
        ZO(I)=Z(I)
    8   Z(I)=ZN(I)
        IF(N/5*5.LT.N) GO TO 100
        WRITE(6,10)T
   10   FORMAT(//F10.3//)
        WRITE(6,20)(Z(I),I=1,IMAX)
   20   FORMAT(10F10.3)
        IF(N.LT.NMAX) GO TO 100
        STOP
        END
```

The following values are taken: $DX = 100$ m, $DT = 1$ s, $FRICT = 0$ (frictionless fluid), $PER = 60$ s, $AMPL = 1$ m, $NMAX = 100$ and a bottom slope of 1%. The water surface elevation values are printed for

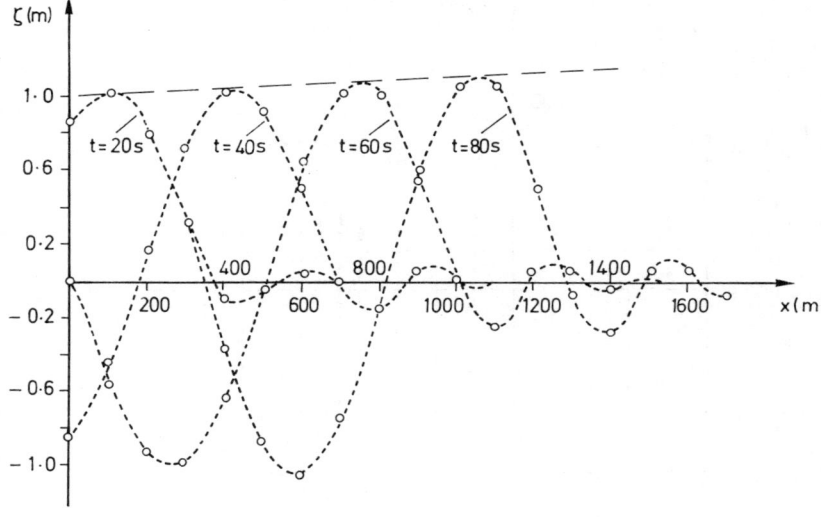

Fig. 7.10

every fifth time step and are graphically presented in Fig. 7.10, showing the wave propagation along the channel. It is interesting to note the numerical dispersion implicit in the difference equations generated by the integration scheme adopted here.

7.6. APPLICATION TO GROUNDWATER FLOW

As an application of the method of finite elements to groundwater flow let us consider the transient response of an unconfined aquifer to a rainfall load.

The aquifer is bounded by a horizontal impermeable base. The left and right boundaries are an impermeable boundary and a water body (river or lake) of constant surface elevation, respectively. The aquifer is shown in Fig. 7.11 with the discretisation in unequal linear elements.

For this special case the Boussinesq equation takes the form:

$$\frac{\partial h}{\partial t} = \frac{k}{2n} \left(\frac{\partial^2 h^2}{\partial x^2} \right) + q \tag{7.101}$$

The adopted values for the permeability and porosity are $K = 50$ m/hr, and $n = 0.4$ respectively. The rainfall intensity corresponding

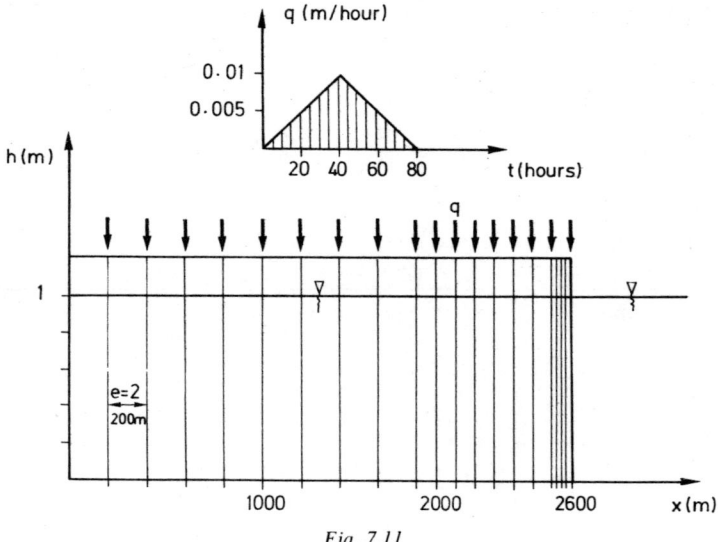

Fig. 7.11

to the source term q is graphically depicted in Fig. 7.11.

If the Galerkin approach is applied the finite element equations deriving from Equation 7.101 have the form:

$$\sum_{e=1}^{M} \left[\int_{(e)} [N]^{(e)} \frac{\partial}{\partial t} \{h\}^{(e)} N_k^{(e)} \, dx + \right.$$

$$\left. + \frac{k}{2n} \int_{(e)} \frac{\partial}{\partial x} [N]^{(e)} \{h^2\}^{(e)} \frac{\partial}{\partial x} N_k^{(e)} \, dx - \int_{(e)} q N_k^{(e)} \, dx \right] = 0$$

(7.102)

An explicit in time integration scheme is used and the time derivative of h is replaced by the forward difference

$$\frac{\partial h}{\partial t} = (h^{n+1} - h^n)/\Delta t \qquad (7.103)$$

where n is the time index, while the h^2 values appearing in the r.h.s. of Equation 7.101 are taken at n time level.

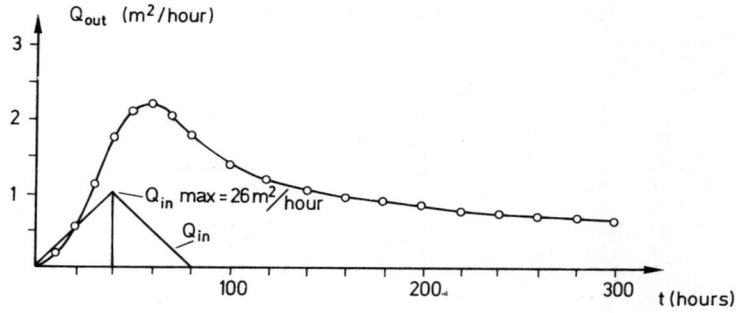

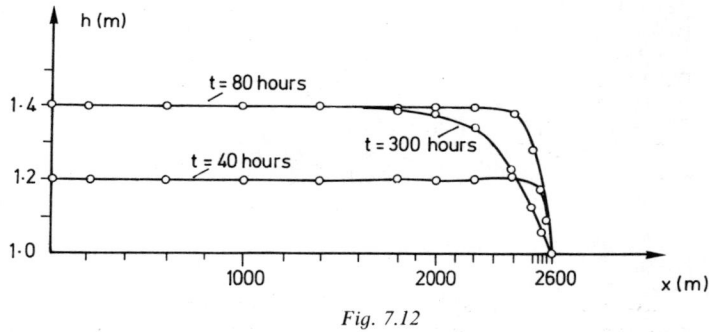

Fig. 7.12

The contributions of the element (*e*) to the global system of equations are,

$$A_{i,i} = \Delta x_i / 3 \tag{7.104}$$

$$A_{i,i+1} = \Delta x_i / 6 \tag{7.105}$$

$$A_{i+1,i} = \Delta x_i / 6 \tag{7.106}$$

$$A_{i+1,i+1} = \Delta x_i / 3 \tag{7.107}$$

$$B_i = (h_i^n/3 + h_{i+1}^n/6)\Delta x_i - (k/2n)\Delta t(h_i^{2n} - h_{i+1}^{2n})/\Delta x_i$$
$$+ q\Delta t \Delta x_i / 2 \tag{7.108}$$

$$B_{i+1} = (h_i^n/6 + h_{i+1}^n/3)\Delta x_i - (k/2n)\Delta t(-h_i^{2n} + h_{i+1}^{2n})/\Delta x_i +$$
$$+ q\Delta t \Delta x_i / 2 \tag{7.109}$$

134 THE METHOD OF FINITE ELEMENTS

The assembly of the elemental contributions and the solution of the simultaneous equations for each time step is programmed in FORTRAN as follows:

```
C     FINITE ELEMENT SOLUTION OF UNCONFINED FLOW
      DIMENSION DX(20),A(21,21),B(21),H(21),HN(21)
      READ(5,1) DT,PERM,POR,QMAX,HO,IMAX
    1 FORMAT(5F7.0,I4)
      IMAX1=IMAX-1
      READ(5,2)(DX(I),I=1,IMAX1)
    2 FORMAT(10F7.0)
      DO 3 I=1,IMAX
      HN(I)=HO
    3 H(I)=HO
      DO 4 I=1,IMAX
      DO 4 J=1,IMAX
    4 A(I,J)=0.
      DO 5 I=1,IMAX1
      A(I,I)=A(I,I)+DX(I)/3
      A(I,I+1)=A(I,I+1)+DX(I)/6
      A(I+1,I)=A(I+1,I)+DX(I)/6
    5 A(I+1,I+1)=A(I+1,I+1)+DX(I)/3
      N=0
      T=0.
  100 T=T+DT
      N=N+1
      IF(N-40)6,6,7
    6 Q=QMAX*N/40
      GO TO 8
    7 IF(N-80) 9,9,10
    9 Q=QMAX*(80-N)/40
      GO TO 8
   10 Q=0.
    8 CONTINUE
      DO 12 I=1,IMAX
   12 B(I)=0.
      DO 13 I=1,IMAX1
      B(I)=B(I)+(H(I)/3+H(I+1)/6)*DX(I)-PERM/2/POR
     1 *DT/DX(I)
     1 *(H(I)**2-H(I+1)**2)+Q*DX(I)/2*DT
   13 B(I+1)=B(I+1)+(H(I)/6+H(I+1)/3)*DX(I)-PERM/2/
     1 POR*DT/DX(I)
     1 *(-H(I)**2+H(I+1)**2)+Q*DX(I)/2*DT
      ITER=0
   99 ITER=ITER+1
      IF(ITER.GT.100) STOP 1
      DIFMX=0.
      DO 14 I=2,IMAX1
      TEMP=HN(I)
      HN(I)=(B(I)-A(I,I+1)*HN(I+1)-A(I,I-1)
     1 *HN(I-1))/A(I,I)
      DIF=ABS(TEMP-HN(I))
      IF(DIF.GT.DIFMX)DIFMX=DIF
```

```
      14   CONTINUE
           IF(DIFMX.GT..0001) GO TO 99
           HN(1)=HN(2)
           DO 15 I=1,IMAX
      15   H(I)=HN(I)
           IF(N/10*10.LT.N) GO TO 100
           QOUT=PERM*(HN(IMAX1)-HN(IMAX))*POR*HO
           WRITE(6,18) T,QOUT
      18   FORMAT(2F10.3)
           WRITE(6,20)(H(I),I=1,IMAX)
      20   FORMAT(21F5.2)
           IF(N.LT.365) GO TO 100
           STOP
           END
```

The data values are: PERM = 50 m/hr, DT = 1 hr, POR = 0.4, QMAX = 0.01 m/hr, HO = 1 m, IMAX = 21. The program computes the outflow discharge QOUT of the aquifer towards the water body by means of the Dupuit formula:

$$Q_{\text{out}} = \frac{Vh}{n} = -\frac{k}{n}\frac{\partial h}{\partial x} h \qquad (7.110)$$

The development of the outflow is given in Fig. 7.12 together with the water surface variation inside the aquifer body.

REFERENCES

1. M. Abbott, *An Introduction to the Method of Characteristics*, Thames & Hudson (1966)
2. M. Abbott, *Computational Hydraulics*, Pitman (1979)
3. F. Abraham and W. Tiller, *An Introduction to Computer Simulation in Applied Science*, IBM Data Proc. Div., Palo Alto Science Center (1970)
4. A. Biswas, *Systems Approach to Water Management*, McGraw-Hill (1976)
5. C. A. Brebbia and H. Tottenham, *Variational Methods in Engineering*, Southampton University Press (1973)
6. T. Chung, *Finite Elements Analysis in Fluid Dynamics*, McGraw-Hill (1978)
7. S. Conte and C. de Boor, *Elementary Numerical Analysis*, McGraw-Hill (1972)
8. J. Connor and C. A. Brebbia, *Finite Element Techniques for Fluid Flow*, Newnes-Butterworths (1976)
9. B. Finlayson, *The Method of Weighted Residuals and Variational Principles*, Academic Press (1972)
10. R. H. Gallagher (Ed.), *Finite Element in Fluids*, Vol. I, II, Wiley (1975)
11. R. Hamming, *Introduction to Applied Numerical Analysis*, McGraw-Hill (1971)
12. F. Henderson, *Open Channel Flow*, Macmillan (1966)
13. K. Huebner, *The Finite Element Method for Engineers*, Wiley Interscience (1975)
14. E. Isaacson and H. Keller, *Analysis of Numerical Methods*, Wiley (1966)
15. G. Marchuck, *Numerical Methods in Weather Prediction*, Academic Press (1974)
16. D. McGracken and W. Dorn, *Numerical Methods and FORTRAN Programming*, Wiley (1964)

17. K. Medearis, *Numerical-Computer Methods for Engineers and Physical Scientists*, KMA (1974)
18. A. Mitchell and D. Griffiths, *The Finite Differences Method in Partial Differential Equations*, Wiley (1980)
19. S. Nakamura, *Computational Methods in Engineering and Science*, Wiley (1977)
20. A. Raudkivi and R. Callander, *Advanced Fluid Mechanics*, Arnold (1975)
21. I. Remson, *Numerical Methods in Subsurface Hydrology*, Wiley (1972)
22. R. Richtmuer and K. Morton, *Difference Methods for Initial Value Problems*, Interscience (1975)
23. P. Roache, *Computational Fluid Dynamics*, Hermosa Publications (1976)
24. G. Smith, *Numerical Solution of Partial Differential Equations*, Oxford University Press (1969)
25. G. Strang and G. Fix, *An Analysis of the Finite Element Method*, Prentice Hall (1973)
26. N. Yanenko, *Méthode à Pas Fractionnaires*, Armand Colin (1968)
27. O. Zienkiewicz, *The Finite Element Method in Engineering Science*, McGraw-Hill (1971)
28. W. Gray and G. Pinder, 'On the relationship between the FE and FD method', *Int. J. Num. Methods Engng*, Vol. 12, No. 9 (1976)
29. Finlayson and Scriven, 'The method of weighted residuals — a review', *Appl. Mech. Rev.*, Vol. 19, No. 9 (1967) pp. 735-748

Index

Advection 67, 70, 99, 102
Algebraic system 11, 21, 51, 111
Algorithm 2
Aliasing 19
Aquifer 89

Back substitution 14
Backward difference 5, 22, 29, 39, 103
Backwater curve 80
Base functions 111
Boundary conditions 24, 29, 30, 45, 57, 66
Boussinesq equation 89, 93, 131

Canonical form 27
Carré method 44
Central differences 22, 29, 39, 41, 43
Characteristic curves 36, 57, 60
Cheng-Allen scheme 31
Chézy coefficient 65
Choleski method 13
Colour equation 37
Confined flow 89
Conservative quantity 99
Consistency, of finite difference scheme 23
Convergence 14, 16, 17, 32
Courant number 39, 76
Crank–Nicolson scheme 32, 34
Critical flow, 81, 82
Curved boundaries 45

Darcy coefficient 49, 54
Darcy law 88, 90
Difference equation 23, 35
Diffusion 28, 99
Dirichlet boundary condition 43
Discharge 25, 48, 52, 66, 93, 135
Discrete form 18
Discretisation 21, 24, 105, 113
Discriminant 27
Dispersion 100
Duffort–Frankel scheme 31
Dupuit approximation 89, 135

Elasticity modulus 56
Elastic wave speed 55
Elliptic equation 28, 43, 88, 91
Energy conservation 27, 48, 51
Energy line 49, 65, 81
Equilibrium 27, 28
Error function 102
Euler equation 107
Evacuation of tank 25
Explicit finite differences 23, 29, 68, 76, 94, 128

Fick's law 99
Finite differences method 21, 22, 23, 24
Finite element method 105
Flood hydrograph 66
Flood routing 67
Forward difference 5, 22, 25, 29, 39, 132
Fourier finite series 18, 20
Free surface boundary 88
Friction coefficient 49, 65, 127
Fromm scheme 43, 103
Functional 107

Galerkin 111, 112, 122, 127, 132
Gauss–Jacobi method 15
Gauss–Jordan method 12
Gauss–Seidel method 16, 33, 44, 46
Grid 34
Groundwater flow 88, 131

Hardy-Cross method 51
Harmonic components 18, 19
Head loss 48, 49
Heat equation 28
Hydraulic radius 65
Hyperbolic equation 27, 67, 101

Impermeable boundary 88
Implicit finite differences 23, 29, 69
Initial conditions 24, 25, 30, 37
Instability 37
Integration by parts 122
Interpolation 3, 61
Iteration 12, 15, 34, 73

INDEX

Labee method 51
Lagrange polynomials 7
Laplace equation 27, 47
Lax-Wendroff scheme 40, 41, 42, 69
Leap frog scheme 41
Least squares 3

Manning formula 65
Mass continuity 27, 48, 65, 81, 88
Minimisation 4, 111
Model, mathematical 1
Molecular diffusion 99

Newmann boundary condition 43
Newton formula 5, 8
Node 21, 114
Normal depth 81
Numerical approximation 3
Numerical dissipation–diffusion 27, 48, 56, 65, 81, 88
Numerical dispersion 38, 42, 75

Open channel flow 65, 80
Ordinary differential equations 21, 25, 108, 118
Orthogonality 18, 19

Parabolic equation 27, 28, 67, 90, 101
Partial differential equations 21, 27
Péclet number 102
Permeability coefficient 81, 88, 89
Phase error 37
Piecewise approximation 114
Pipe flow 48, 55
Pipe network 48
Poisson equation 43, 96
Pressure 55
Pollutant 99
Potential velocity 46, 88, 89
Propagation 28
Pump 49

Relaxation method 17, 44
Reflection boundary condition 67, 77, 101
Reservoir boundary condition 88

Resistance, hydraulic 49
Residual 111, 112
Reynolds number 49
Ritz method 108, 118

Seepage surface 88
Seepage velocity 88
Shape function 106, 113, 114
Shear stress 3, 57, 127
Simpson formula 10
Spectral radius 15, 16, 17
St. Venant equations 126
Stability 23, 31, 39, 40, 95, 101
Staggered grid 58

Taylor series 22, 40, 45
Telegraphy equation 127
Thomas's method 14
Time step 25, 29
Transient 27
Translation operator 5
Transmission (free) boundary condition 67, 101
Trapezoidal rule 10
Tree-shape network 50
Triangular matrix 12, 14
Triangular element 115
Turbulent diffusion 100

Unconfined flow 89
Undetermined coefficients 9, 111

Variable step method 83
Variational method 105, 106
Velocity distribution 3
Velocity, mean 65
Viscosity 29

Water hammer 55
Wave equation 28, 36, 37
Wave (flood) 65, 71
Wave (tidal) 65, 77, 126
Wave propagation 67
Wave speed 28, 71, 78
Weighted residuals method 110